THE EXPRESSIONIST SURFACE

CONTEMPORARY ART IN PLASTER

BARBARA C. MATILSKY

June 9 - August 26, 1990

THE QUEENS MUSEUM
New York

THE QUEENS MUSEUM STAFF

ADMINISTRATION
Steven Klindt, Executive Director
Beth Henriques, Assistant Director
Barbara Sperber, Director of Administration
Connie M. Cullen, Public Information Officer
Louis Acquavita, Facility Manager
Frances Hyatt, Executive Secretary
Doretha Moorer, Administrative Assistant
Christine Shepard, Receptionist/Typist
John Poulson, Museum Maintenance

DEVELOPMENT
David Rodriguez, Director of Development
Kitty Schwartz, Development Officer
David Freilach, Development Officer/Grants Writer
Grace Woods, Secretary

PROGRAMS
Barbara C. Matilsky, Curator
Marc H. Miller, Curator
Dawn A. Giegerich, Registrar
Phyllis Bilick, Assistant Curator/Photographer
Sunday Englis, Conservator
Joana A. Vicioso, Secretary

EDUCATION
Sharon Vatsky, Curator of Education
Lorraine Klagsbrun, Assistant Curator of Education
Miriam Brumer, Museum Educator
Tom Brydelsky, Museum Educator
Pearl Rosen, Special Education Coordinator
Shelly Wright, Secretary

SECURITY
Anthony Kemper, Chief of Security
Willie Ed Guy, Assistant Chief
Andrew Strowder, Assistant Chief
Richard Simmons
Oneather Folks
Eric Lima
Judith Mendez

Support for this exhibition has been provided by The National Endowment for the Arts, a Federal agency, and the New York State Council on the Arts.

queens museum

The Queens Museum is housed in the New York City Building, which is owned by the City of New York, and its operation is supported in part with public funds provided by the New York City Department of Cultural Affairs, and the Office of the Queens Borough President Claire Shulman. Additional support is provided with public funds from the New York State Legislature and New York State Council on the Arts.

Library of Congress Catalog Card Number 90-60494

ISBN 0-9604514-2-0

DESIGN Karen Davidson, Exhibition Catalogues, New York

PRINTING Studley Press, Dalton, Massachusetts

COVER PHOTOGRAPH: Rudolph Serra, *Vapa* (detail), 1989, Hydrocal

TABLE OF CONTENTS

ACKNOWLEDGEMENTS

The Expressionist Surface is the first exhibition to document the innovative uses of plaster in contemporary art. Traditionally, plaster was considered a support material, used primarily for casting bronze sculptures. Only recently have artists discovered its expressive potential and fluidity which directly translates the artist's process. Using plaster to create autonomous works of art is a new and important development in the history of this ancient material.

Dr. Barbara C. Matilsky, Curator, conceived and worked on this exhibition for over two years, visiting artists' studios and discussing with them works in progress. Her enthusiasm and commitment to the project, and the artists' cooperation and support were essential to the exhibition's success. The contributions made by all of the artists to the show are especially appreciated. We thank Claude Fixler, Grace Knowlton, Ray Rapp, and Jonathan Silver for their time and energy while creating sculpture installations specifically for the show.

The Queens Museum appreciates the generosity of the private collectors who loaned work to the exhibition: Ms. Paule Anglim, Mr. Joseph Cavaliero, Mr. & Mrs. Robert Hudson, and Jack and Harriet Stievelman. The Charles Cowles Gallery, the Sidney Janis Gallery, OK Harris and Ann Kohs & Associates were instrumental in facilitating key loans.

Suggestions regarding the contents of the catalogue were contributed by Gail Stavitsky, Michelle Rutman, Alan Moore, and Marc H. Miller, Curator at The Queens Museum. In addition, many staff members at The Queens Museum were important resources at various stages of the project: Ileen Sheppard, former Director of Exhibitions, Dawn A. Giegerich, Registrar, Connie M. Cullen, Public Information Officer, and Sunday Englis, Conservator, who was responsible for researching and compiling the catalogue glossary. Phyllis Bilick, Assistant Curator, contributed her expertise in many capacities, from advice on object selection to the photographing of works of art. A special note of thanks to Janet Schneider, former Executive Director of The Queens Museum, who first recognized the importance of this exhibition and offered encouragement during the initial stages of its development.

The Queens Museum is grateful to the National Endowment for the Arts and the New York State Council on the Arts for funding the exhibition.

Steven Klindt
Executive Director

PLASTER TRANSFORMED

Artists have only recently begun to recognize the aesthetic and expressive possibilities inherent in plaster. An ancient material, plaster was traditionally valued for its functional and supporting role in conjunction with other materials and techniques. Painters used it as a ground for frescoes and panel painting, and sculptors employed it to replicate the human body and cast objects into other, "finer" materials, usually bronze. Plaster is now considered a primary material for independent works of art. Contemporary artists find plaster's fluidity a perfect vehicle for interpreting emotion, creating new forms or updating traditional approaches. They have liberated plaster from its mundane, subservient status, simultaneously forging a new identity for the material and for themselves as artists.

The Expressionist Surface focuses on sixteen established and emerging artists who work almost exclusively in plaster. Their paintings, sculptures and site-specific installations range from representational to abstract, and are often characterized by multi-faceted textured surfaces, relentless energy and a deep subjective vision. Plaster is used by many artists to merge painting and sculpture, adding another layer to the physical and psychological composition of the work. By manipulating and painting the material to reinforce the drama and urgency of their themes, the artist's "gesture" or mark remains visible.

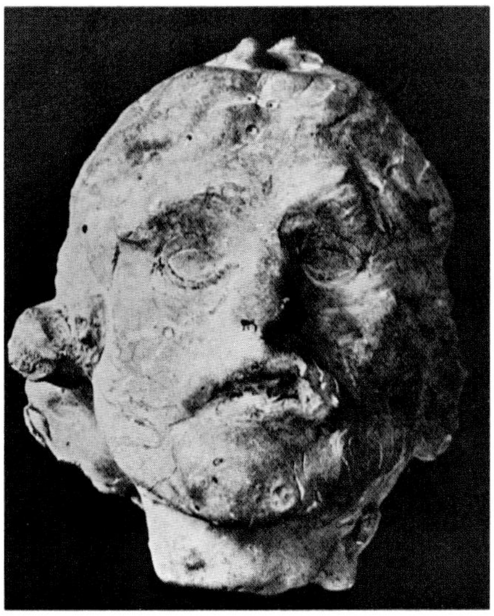

Fig. 1. August Rodin, Study of the *Head of St. John Baptist*, 1877-78, dipped plaster, Musée Rodin, Paris (from *Rodin Rediscovered*, ed. Albert E. Elsen, National Gallery of Art, Washington, D.C., 1981, p. 138)

This orientation differs from many artists who emphasize plaster's smoothness to achieve an effect of aloof elegance and "cool" emotion. The distinction between Auguste Rodin's sculptures with their molten, erupting surfaces and Constantin Brancusi's streamlined aesthetic is a useful, though simplified, point of comparison and reference. The artists in this show also differ from contemporary sculptors who continue plaster's use for replication as they redefine conceptions of art: Tom Otterness creates multiples to popularize art, thereby democratizing the marketplace; Edward Allington appropriates miniature replicas of Venus de Milo to underscore the debasement of classical images; Allan McCollum casts hundreds of objects to express the loss of art's originality and its subsequent reduction to a neutral entity.

The artists in this exhibition express a wide spectrum of feelings, disclosing art's full physical, spiritual, and emotional dimensions. The themes in their works embody ideas of flux — emergence, disintegration, renewal and transformation. Whether interpreted through stories drawn from myth or the Bible, references to the cycles, geometries and essences of nature, or to the mutabilities of the human condition, the resolution of chaos and order is a common thread running through the work.

The painting and sculpture in *The Expressionist Surface* may be characterized as romantic. This romanticism, however, is not defined by nostalgia, sentimentality or yearning, but by an identification with tradition, and the painful realization that life's complexity obliterates the purity of ideal values. The knowledge results not in cynicism or criticism, but in a process of self-discovery whereby art affirms a place within a society caught in the throes of radical change. These artists accept many of the philosophical premises of Abstract Expressionism as they were defined during the 1950s, particularly the belief that creating art defines one's authenticity and a reassuring niche in life's existential void. Abstract Expressionism emerged during a time of great ambivalence: post-war economic expansion and optimism coincided with the fears generated by Communism and the recently unleashed atomic bomb. Contemporary artists face less cataclysmic, subtler and more insidious circumstances as they struggle for identity in a fast-paced, post-industrial world motivated by status and wealth. Conformity and global homogeneity are perceived as constant threats to artists who reject the current fashion for consumer and media-oriented art. In response, artists seek alternatives as they reinterpret the tradition of art to reflect the complexities of late twentieth-century life. Working in plaster is just one of the many options open to artists for reevaluation, experimentation and synthesis of ideas.

The Nature of Plaster and its Interpretation

Most people are familiar with plaster as a building material in the form of joint compound and sheetrock for erecting walls. Ordinary plaster, named Plaster of Paris after the countryside around the French capital where the material was mined, is derived from the mineral gypsum. Occurring naturally in rock form, it is pulverized and heated in a kiln to evaporate three-quarters of its water content. The resulting powder is then ready to be rehydrated into the desired form. (A variety of plasters are commercially available. For a listing of the most popular types and their properties, see glossary on page 48.) Mixed with water, plaster's viscosity ranges from that of syrup to peanut butter. At this point, the artist can add a variety of ingredients: sand, perlite, fiberglass, polymers, and pigments. Artists must work rapidly if the type of plaster they select dries quickly. Through trial and error, they must also determine the right mixture for obtaining the desired effect, unless prepared plasters like joint compound or plaster bandages are used.

Compared to other natural materials traditionally used for making sculpture — stone, wood, clay — plaster is the most versatile. It can be modelled, poured and thrown when wet, carved dry, cast, and reconfigured into complex combinations of forms. Artists sometimes embed objects into the material — a white composition of unusual pattern and mystery results. By contrast, marble, granite and sandstone can only be carved, a subtractive process of eliminating sections from the rock. Sculptors also carve wood by chiselling, and more recently, by motor-powered chain saw. Wood can also be constructed, as individual pieces of the material are assembled and integrated into a unified composition. The artist's ease or difficulty in translating her vision is determined by the variety of stone or species of wood. Clay, like plaster, is malleable, and the artist models by building up and kneading it into a desired shape; clay is also carved, usually with a knife. Although artists employ clay in casting, it is not as fluid as plaster.

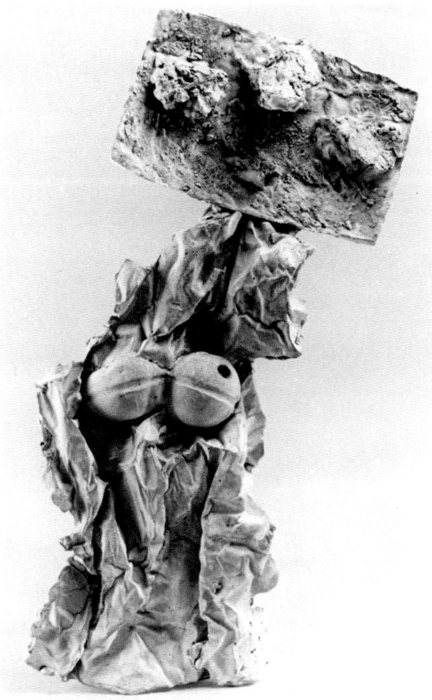

Fig. 2. Pablo Picasso, *Personnage*, 1933, Musée Picasso, Paris. Photo © R.M.N.

Plaster is inherently luminous, and contrasting with the shadowed hollows of a textured surface, the reflected light contributes to the dramatic aspect of a work. The glowing whiteness induces an apparitional, dream-like effect which, in the figurative works of Christopher Cairns, Claude Fixler, George Segal and Jonathan Silver, conveys psychological and metaphysical realities.

Plaster's greatest disadvantage is its weakness, breaking by accident and weathering when exposed to the elements. For this reason, most artists in the past shunned plaster for finished works of art. New varieties like Hydrocal are denser and stronger than regular plaster. Structolite and Design Cast can even withstand extremes of climate. Grace Knowlton and Ed Smith regularly exhibit sculptures made from these newer materials outdoors. If properly handled and exhibited, plaster is a relatively stable material for works of art. The preservation of painted plaster sculptures dating from 7000 B.C. testify that, given proper temperature and humidity, the material endures.

Not stability but fluidity is plaster's primary asset. Its capability of transformation exposes artists to the possibilities of spontaneity in their process. As a powder, it is formless; the solidified result looks nothing like the material in its virgin state. Artists find this aspect challenging. Chance and accident can be welcome determinants of form, aligning the art work with the realities of life. Dada artists of the early twentieth century first championed the concept of chance, and used chance procedures to activitate their compositions and expand the boundaries of their art. Process Artists of the mid-1960s and '70s broadened this approach by flinging, scattering, pouring and throwing materials, allowing gravity, velocity and the materials themselves to determine the final effect. Plaster is similarly used by Ray Rapp, Rudolph Serra and Ed Smith who interpret the flowing nature of the material in their work.

Artists have traditionally used plaster to reproduce other objects and human figures through casting — the process of making a mold and obtaining a positive impression. Contemporary artists, while continuing to cast, have gone beyond the naturalism of pure reproduction. After first modelling his figures, George Segal began casting live models and juxtaposing the hollow shells with architectural fragments. His work forecast the current trend in installation art. Christopher Cairns, Claude Fixler and Jonathan Silver cast clay and plaster figures, innovatively composing forms and/or creating environmental tableaux to mirror aspects of reality. The harmonies and energies of nature are revealed by Peter Agostini and Grace Knowlton who cast synthetic and organic materials and objects into abstract shapes. Ray Rapp and Rudolph Serra cast wood forms that refer to pure geometry, land formations and architecture.

Plaster provides artists with the means to merge painting and sculpture. As painters, Susan Laufer and Robert Van Vranken pursue rich, palpable textures. These artists build up, incise, gouge and sand the plaster surface, injecting dimension into an otherwise flat picture plane. To add color, artists mix pigments into wet plaster, directly apply paint to the material after it hardens, or combine approaches. Sculptors seeking to enhance the surface and volume of their work through color find plaster accepts paint easily without compromising the nature of the material. Manuel Neri, Elfi Schuselka and George Segal paint directly on plaster, while Anita Curtis Glesta, Barry Goldberg, Grace Knowlton and Ed Smith impregnate plaster with pigment, making sculpture and painting simultaneously, refining tonality after the surface dries. The majority of these sculptors began their careers as painters. Plaster enabled them to make the transition and to integrate many of painting's formal concerns into their work.

Plaster lends itself to experimentation because of the diverse effects that artists can achieve. Its immediacy and fluidity also encourages both conceptual and formal freedom. Purchased from any hardware store, it is accessible and, most importantly, inexpensive. Artists can afford to experiment and discard unsatisfactory results with less hesitation.

Plaster Develops an Independent Identity

The earliest examples of art in plaster date to 7,000 B.C. In the prehistoric city of Jericho, human skulls were modelled with pigmented plaster in an early form of portraiture. Plaster's association with funerary ritual continued in ancient Egypt where brightly colored mummy masks were found in royal tombs dating from 2,600 B.C. Until relatively recently, plaster death masks were made of notable individuals and later used as reference for creating portraits and historical monuments. In painting, plaster appears in Minoan frescoes (1,700 B.C.), housed in the Palace of Knossos on Crete. True fresco painting,

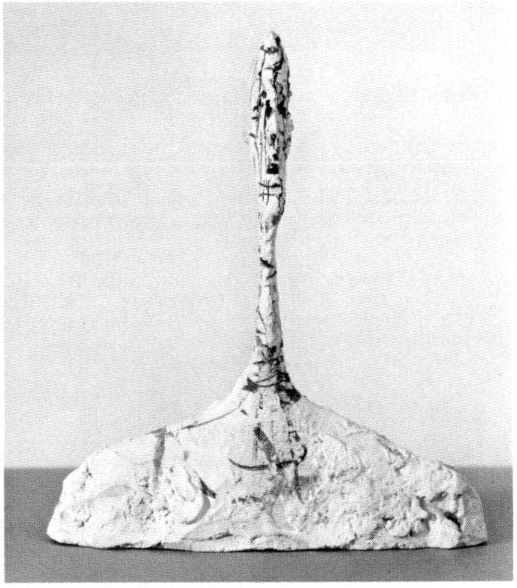

Fig. 3. Alberto Giacometti, *Bust of a Man* (c. 1950), painted plaster, 7 3/4 x 6 5/8 x 2 1/8 inches. Collection, The Museum of Modern Art, New York. Ruth Vollmer Bequest.

the process of using wet plaster as a ground for the application of pigment, has been used to decorate walls for centuries. In these works, the material was rarely used expressively or made visible as an element in the painting's composition.

It was during the nineteenth century when plaster was most widely used, especially at art academies in Europe and the United States. Students of sculpture mastered techniques of casting, modelling, and carving plaster before tackling marble and bronze. Plaster casts of Greco-Roman and Renaissance masterpieces were exhibited in classrooms where principles of drawing were studied. By copying these reproductions, students learned anatomy, proportion and volume before graduating to life drawing courses and directly studying nude models.

During this period, plaster was used as a surrogate material because it was inexpensive and easily cast. At the Paris salons, plaster sculptures were exhibited by artists seeking commissions in marble or bronze. These sculptures were not considered "finished" works, but models for execution into more costly and enduring materials. Plaster sculptures were also useful for reference. To avoid expensive mistakes, artists like Antonio Canova (1752-1822) measured and transferred dimensions from plaster models into marble, and Jean Antoine Houdon (1741-1828) cast live models to aid in his portraiture. Finally, as an intermediate step in the making of bronze sculpture, plaster was used to make molds from clay, wax or plaster models. In most of these instances, after artists completed their work, the plasters were usually neglected or destroyed.

One of the first sculptors to appreciate plaster's distinct qualities was Auguste Rodin (1840-1917), a student of the academic tradition. By giving his plaster sculptures as gifts to friends and donating them to major museums, he appreciated plaster's autonomy as works of art. Rodin also experimented with the material in his innovative process of dip casting, immersing plaster heads and small figures into liquid plaster for unusual, expressive results (Fig. 1)[1] Rodin relied on plaster to explore complex compositions through the unorthodox method of combining casts of torsos, heads and limbs from different sculptures.

Few early twentieth-century artists escaped the dominating influence of Rodin's art, including Pablo Picasso (1881-1973) and Alberto Giacometti (1901-1966). These two pioneering modernists also recognized plaster's aesthetic and expressive value, and used it to create independent works of art. While Rodin maintained a traditional approach to the human form by casting clay and plaster figures, Picasso cast objects in plaster and assembled them into figures (Fig 2).

Picasso used a plethora of materials unconventionally. During the 1930s, he created many plaster sculptures, and according to the photographer, Brassaï, Picasso did not want them translated into bronzes.[2] In *Personnage* (1933), the artist cast creased paper and a cardboard box to approximate the face and body. Then he freely modelled her features, creating an expressively textural and whimsical effect. Plaster proved ideal for Picasso's metamorphosis of objects and experimental interpretation of the human form. By changing the context of every day items, he also redefined their meaning which revolutionized the centuries-old conception of the human figure in art.

The freedom implicit in Picasso's new sculptural language affected Giacometti, who invented his own surrealist vocabulary. Later in his career, however, he returned to the more traditional approach of modelling. Giacometti carved and modelled clay and plaster, intending to cast the results in bronze. Still, a number of painted plasters dating from the 1940s show the artist recognized their worth as finished works of art.[3] In *Bust of a Man* (1950), the tenuous, fragile plaster communicates the subject's angst (Fig. 3). The shoulder ascends into an attenuated neck and flattened head, and color is abstractly dashed onto the nearly disembodied white silhouette; the work is a statement on humanity's vulnerable condition.

Peter Agostini, Manuel Neri and George Segal, who developed careers within the shadows of Abstract Expressionism in the 1950s, were the first artists to create a large body of work in plaster. They were also among the first sculptors to work exclusively with the material. Neri and Segal used it to revitalize the figure during a period when painting dominated and sculpture was primarily non-representational. Agostini modelled figures throughout his career, and created innovative and lyrical abstractions in plaster. All three artists integrated Abstract Expressionism's gesture and emotion into their plaster sculptures, even as Pop and Minimal artists of the mid-'60s embraced a "cooler" aesthetic inspired by the media and industrial materials.

The majority of art work in *The Expressionist Surface* reflects the pluralism that has characterized the art world since Minimal art. Throughout the 1970s, artists experimented with a variety of materials and processes. Arte Povera and Process Art catalyzed this direction, and the Neo-Expressionist paintings of the 1980s reinjected intense emotion and mythic content into art. For the artists in this exhibition, plaster is ideal for evoking feelings while expressing the nature of the material and their individual approach to the creative process. The section that follows illustrates how artists interpret plaster's unique properties in independent works of art.

1 Albert Elsen, "When the Sculptures were White: Rodin's Work in Plaster," *Rodin Rediscovered* (exhibition catalogue), National Gallery of Art, Washington, D.C., 1981, p. 138.
2. Werner Spies, *Sculpture of Picasso*, New York, 1971, p. 103.
3. Alain Kirili, *Alberto Giacometti, Plâtres Peints* (exhibition catalogue), Galerie Adrien Maeght, Paris, 1984.

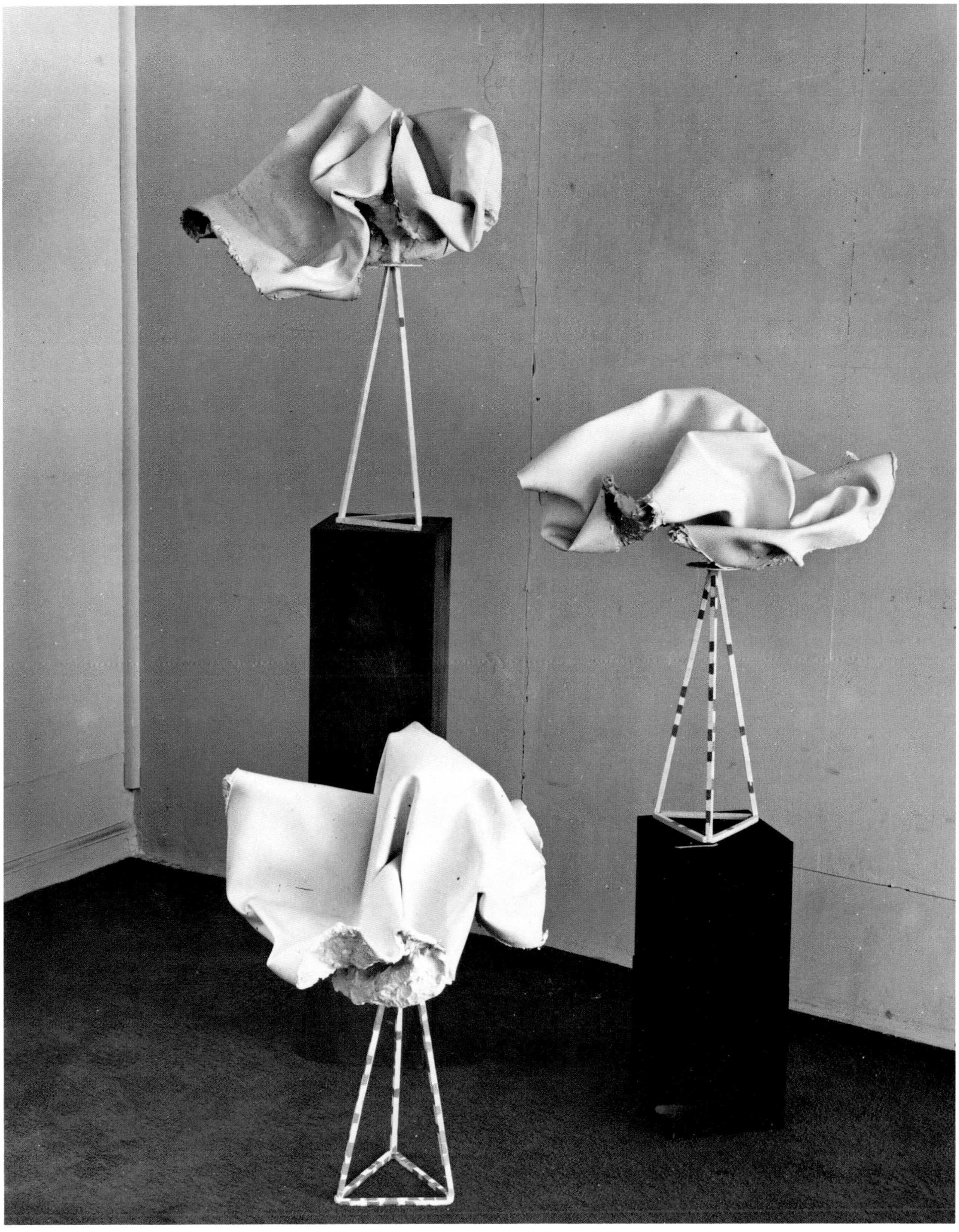

PETER AGOSTINI
Summer Breezes, 1962
Hydrocal and burlap
36 x 30 x 18 inches
36 x 20 x 20 inches
36 x 28 x 20 inches

1913 Born in New York City

Lives and works in New York City

SELECTED AWARDS

1969 National Arts Council Award

1964 Guggenheim Fellowship for Creative Sculpture

1964-1965 Commission for the New York State Exhibition, World's Fair

1964 Brandeis University Creative Arts Award for Sculpture

SELECTED SOLO EXHIBITIONS

1988 Vanderwoude Tananbaum Gallery, New York

1988, 1986, 1985 Bernice Steinbaum Gallery, New York

1976, 1973, 1971 Zabriskie Gallery, New York

1969 Weatherspoon Art Gallery, University of North Carolina, Greensboro

1969, 1968, 1967, 1966, 1963, 1962, 1960 Stephen Radich Gallery, New York

1966, 1965, 1964, 1963, 1962, 1960 Richard Gray Gallery, Chicago

1964 Columbia University, New York

1959 Galerie Grimaud, New York

SELECTED GROUP EXHIBTIONS

1984 "The Third Dimension: Sculpture of the N.Y. School," The Whitney Museum of American Art, New York

1980 "Sculpture In The 70's: The Figure," Pratt Manhattan Center Gallery, New York

1972 American Academy of Arts and Letters Exhibition, New York

1970 "L'Art Vivant aux Etats-unis," Fondation Maeght, St. Paul, Miese, France

1970, 1968, 1966, 1964 "Sculpture Annual," The Whitney Museum of American Art, New York

1967 "American Sculpture of the '60s," Los Angeles County Museum

1963 "Sao Paulo Biennial VII," Sao Paulo, Brazil

1960 "Exhibition of Modern American Painting and Sculpture," University Art Museum, Berkeley, California

PETER AGOSTINI

Peter Agostini has been working with plaster for over fifty years, creating evocative and lyrical sculptures defined by light, line and movement. His achievements in plaster inspired Christopher Cairns and Jonathan Silver whose works are also included in this exhibition. Self-trained, Agostini insists on freedom to experiment. By pushing the boundaries of the material to its furthest limits, he invents forms and introduces unconventional methods. The results run the gamut from figurative to abstract.

In his best-known series of works called *Swells* (1960s), Agostini creates smooth, voluptuous, biomorphic forms from the inner tubes of cars and trucks. Through an ingenious and dangerous technique of inflating, casting and removing of the object, Agostini composes sculptures that embody the tension between expansion and compression; a bursting, explosive energy is sensuously contained within.

Agostini compares making sculpture in plaster to ballet, in that the fluid, graceful contours appear without evidence of its complex staging. His process, based on intuition, control, speed and timing, translates the immediacy of feelings and experiences. E. A. Navaretta portrayed the artist's creation of the sculpture, *Hurricane* (*ARTnews*, May, 1962), by paralleling Agostini's approach to Jackson Pollock's emphasis on action and instinctual feel for materials. This mode of expressionism enables Agostini to interpret the spirit of an idea rather than its literal representation. In *Summer Breezes* (1962, page 11), the essence of movement and the artist's energies merge. Although supported by vertical rods, the sculptures appear to unfurl and float in mid-air; the pristine whiteness of plaster heightens the illusion of weightlessness. This work was shaped by casting plaster and burlap over rubber. Its fluidity of execution is mirrored in the forms which echo the baroque, windswept draperies of Bernini's seventeenth-century marble sculptures.

A similar feeling for movement is suggested in *The Wave* (1963) where Agostini draws with the material in space. Improvisational gesture is transcribed by wrapping plaster soaked in cheesecloth around wire mesh armatures. The artist exposes sections of the structure to delineate the delicate fragmentation within the work. The sense of waves breaking, their ebb and flow, fuses with the artist's own motions. Agostini's abstract form simultaneously traces, veils, and encapsulates space within a thin plaster membrane. In this sculpture, plaster is poetically transformed into patterns of energy and light.

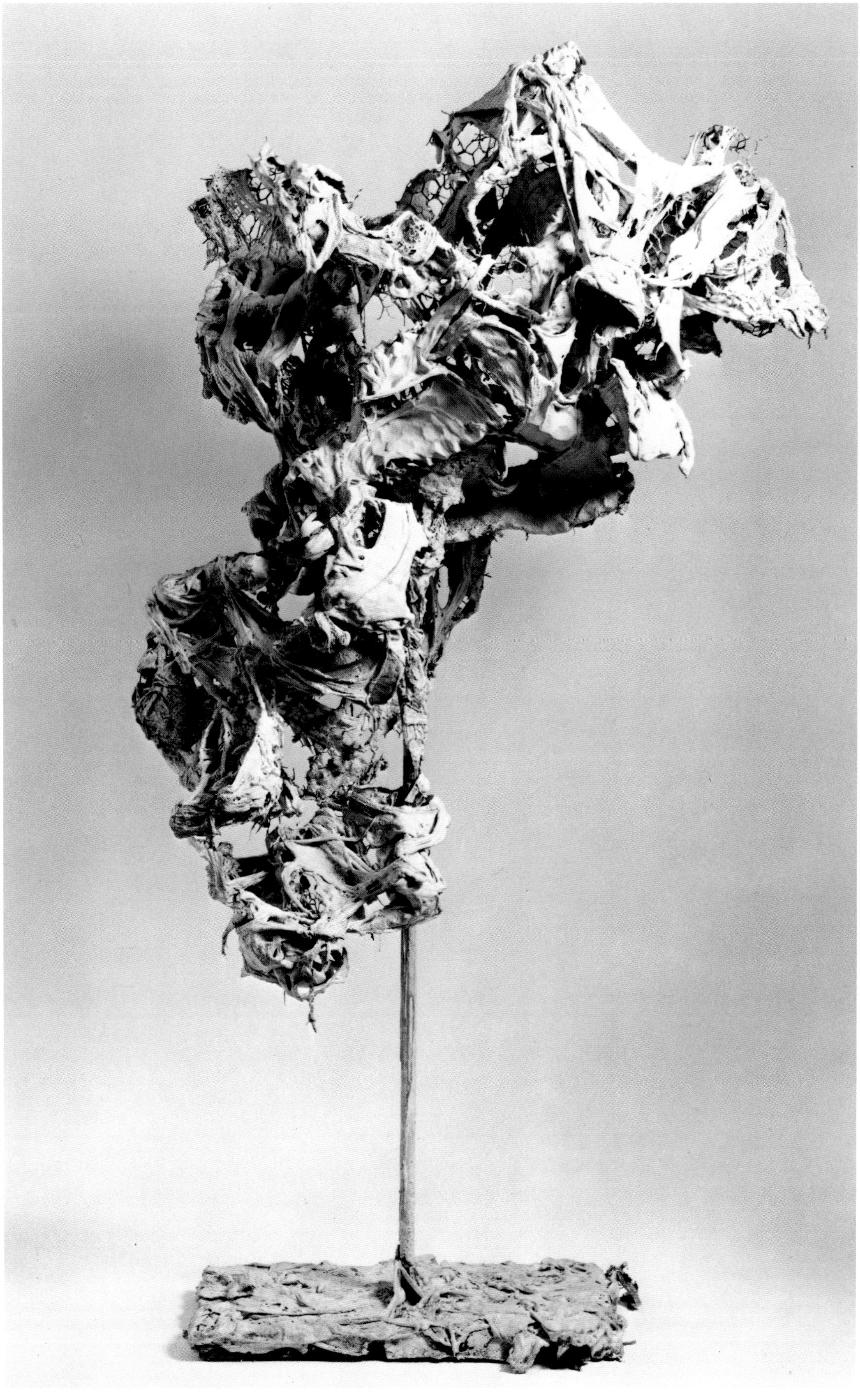

PETER AGOSTINI
The Wave, 1963
Hydrocal, chicken wire and
cheesecloth
62 x 44 x 22 inches

MANUEL NERI

Manuel Neri uses plaster to express the vitality and vulnerability of the female figure. His women always seem open to our penetrating view. Still, no matter how ravaged or distraught they appear, the figures always retain a refined integrity. Neri began working with plaster in the 1950s. Inspired by San Francisco Bay area artists Richard Diebenkorn, Nathan Oliveira, Elmer Bischoff and David Park, Neri discovered the female figure as an expressive agent of emotion.

In *Shrouded Figure* (1961), Neri strikes a balance between classical restraint and expressionistic handling. In a spirited cross between sculptures from the Parthenon and the Hellenistic statuette known as the *Baker Dancer* (200 B.C.), Neri creates a combination of elegance, harmony and intentional fragmentation. The head and arm are cocooned in a single volume of plaster, meeting the amputated leg area in one smooth sweep. Neri unveils a section of the head by chiselling into the plaster to excavate a layer of grey paint. He further activates the surface by bold painting, fluting and slashing. The intimation of what is hidden or missing is as expressive as what exists. Neri explicitly exposes the figure's sexuality while dramatizing her helplessness and incapacity to ward off voyeuristic glances.

Neri evokes a similar sense of helplessness in *The Seated Girl* (1964, page 16). She is posed uneasily, stretched and precariously balanced on a wooden frame. Loosely painted in flaming yellows and reds, she is startlingly unnaturalistic. There is something mask-like about her face, yet the figure is decidedly human. We empathize with her awkwardness, missing pieces and scraped "skin." The spontaneity with which Neri attacks the figure conjures Willem de Kooning's *Women* from the 1950s. The same violently additive and subtractive process drives both artists. Drips and splattered pigment settle in the aftermath.

In a recent sculpture, *Catun Series II* (1986-88), Neri's figure bears a distinct expressive and stylistic affinity to Edgar Degas' bronze *Young Dancer* (1880-81). The same grace, tentativeness and ambiguity define both works. While Degas' figure is restrained, Neri's woman, with broken fragments roped together, reflects his unleashed feelings. Both figures are survivors of man's analysis. Like Degas, Neri's attitude towards women remains difficult to interpret. He chips away at the plaster's wholeness; still, a fragile beauty emerges from the pieces, heightening the work's power and complexity.

1930 Born in Sanger, California

Lives and works in Benicia, California

Professor of Art, University of California, Davis

1957-1959 California School of Fine Arts, San Francisco

1952-1957 California School of Arts and Crafts, Oakland

1951-1952 University of California, Berkeley

SELECTED AWARDS

1985 San Francisco Arts Commission, Award of Honor for Outstanding Achievement in Sculpture

1982 American Academy and Institute of Arts and Letters, Academy-Institute Award in Art

1980 National Endowment for the Arts (Individual Artist Grant)

1979 Guggenheim Fellowship

SELECTED SOLO EXHIBITIONS

1989, 1986, 1982, 1981 Charles Cowles Gallery, New York

1989 "Manuel Neri: Plaster," San Francisco Museum of Modern Art

1987 San Antonio Art Institute, Texas

1984, 1983 John Berggruen Gallery, Washington, D.C.

1981 Seattle Art Museum, Seattle, Washington

1979 Gallery Paule Anglim, San Francisco

1971 San Francisco Museum of Art

1969 Louisiana State University, Baton Rouge

SELECTED GROUP EXHIBITIONS

1990 "Bay Area Figurative Art 1950-65," San Francisco Museum of Modern Art

1986 "California Sculpture: 1959-1980," San Francisco Museum of Modern Art

1984 "American Sculpture: Three Decades," Seattle Art Museum

1976 "California Painting and Sculpture: The Modern Era," San Francisco Museum of Art

1970 "Annual Exhibition: Contemporary American Sculpture," Whitney Museum of American Art

1967 "Funk Art," University Art Museum, Berkeley, California

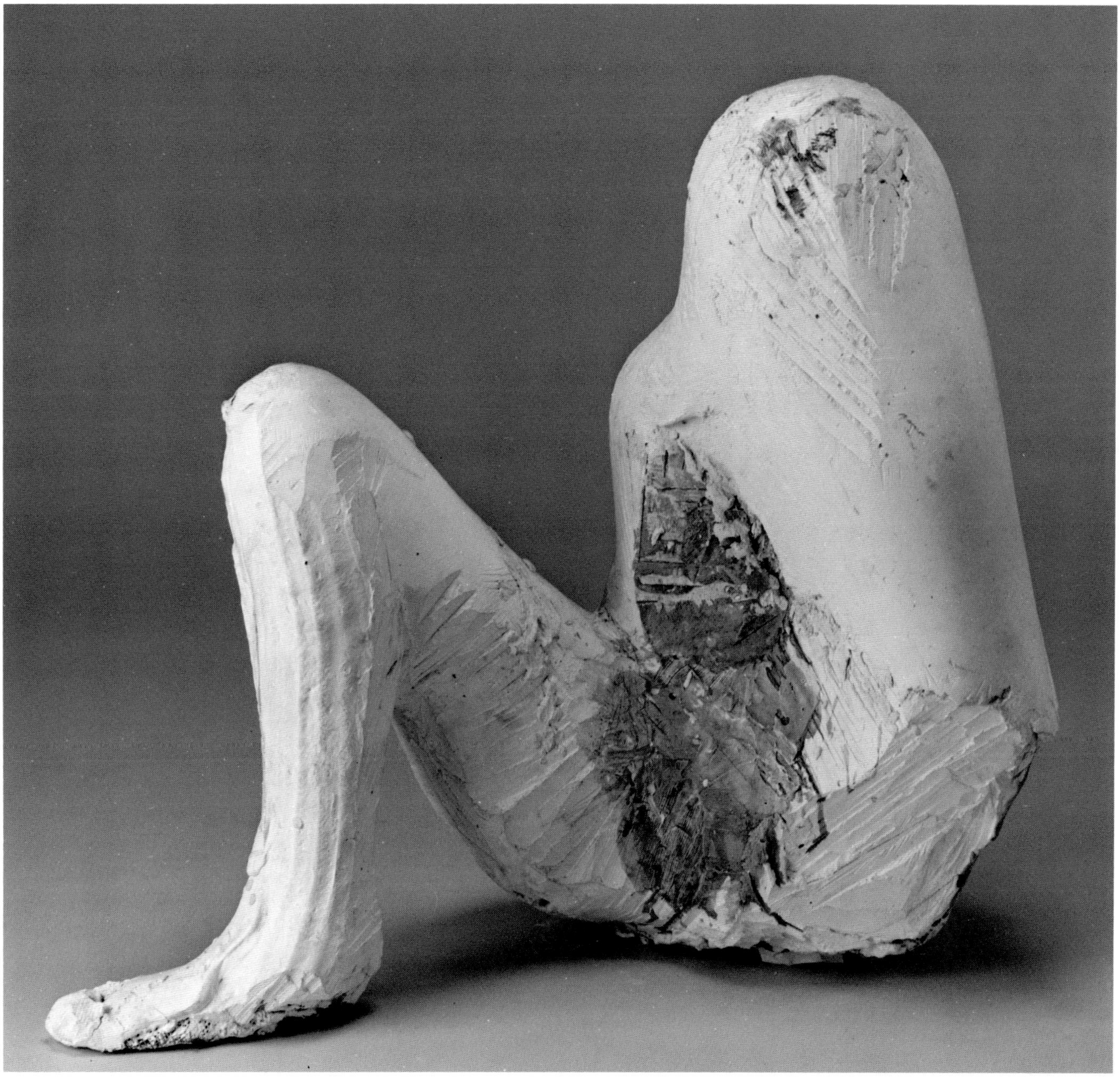

MANUEL NERI
Shrouded Figure, 1961
Plaster, pigment, aluminum, burlap
28 x 32 x 15 inches

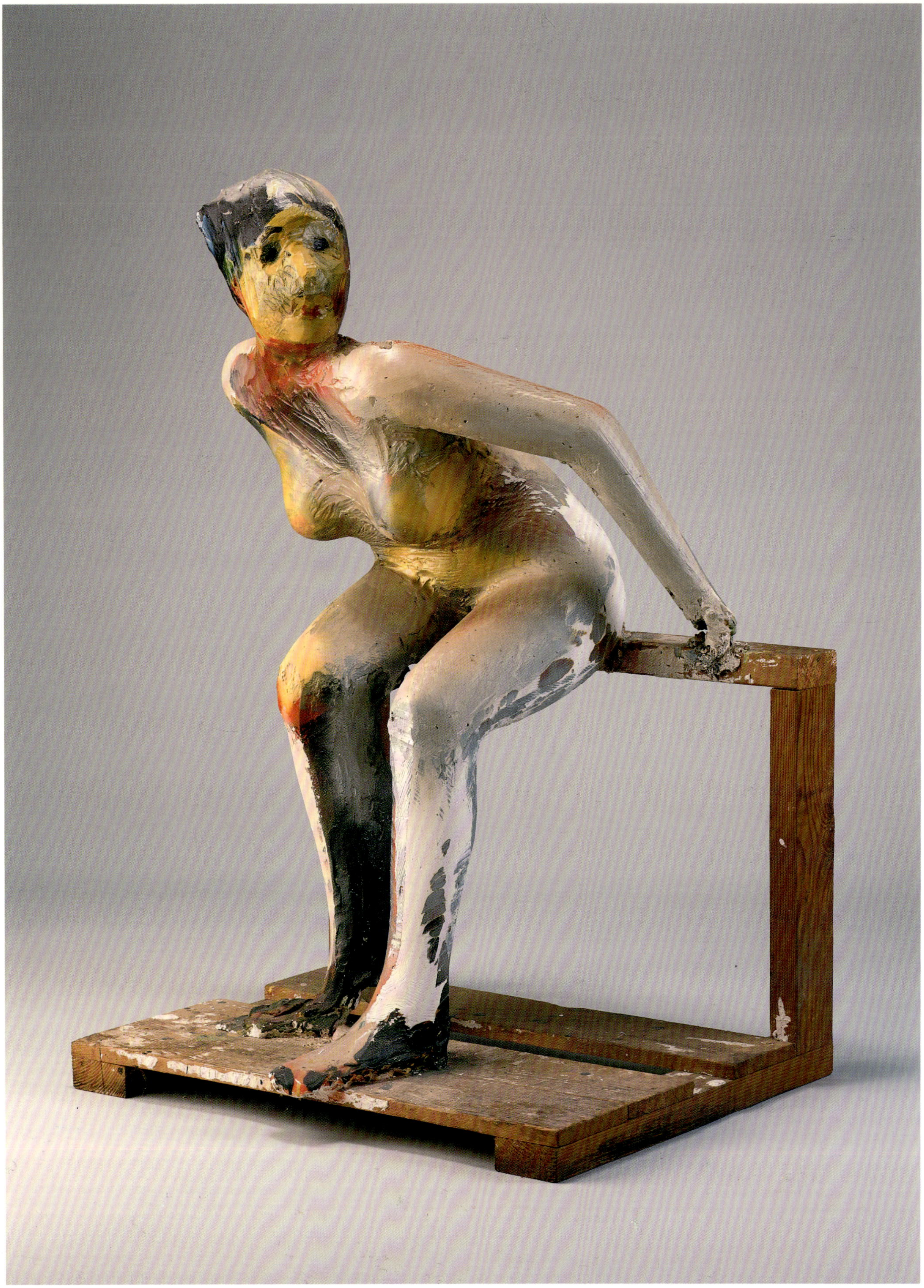

MANUEL NERI
Seated Girl, 1964
Plaster, enamel, wood
44 x 29 1/2 x 27 1/2 inches

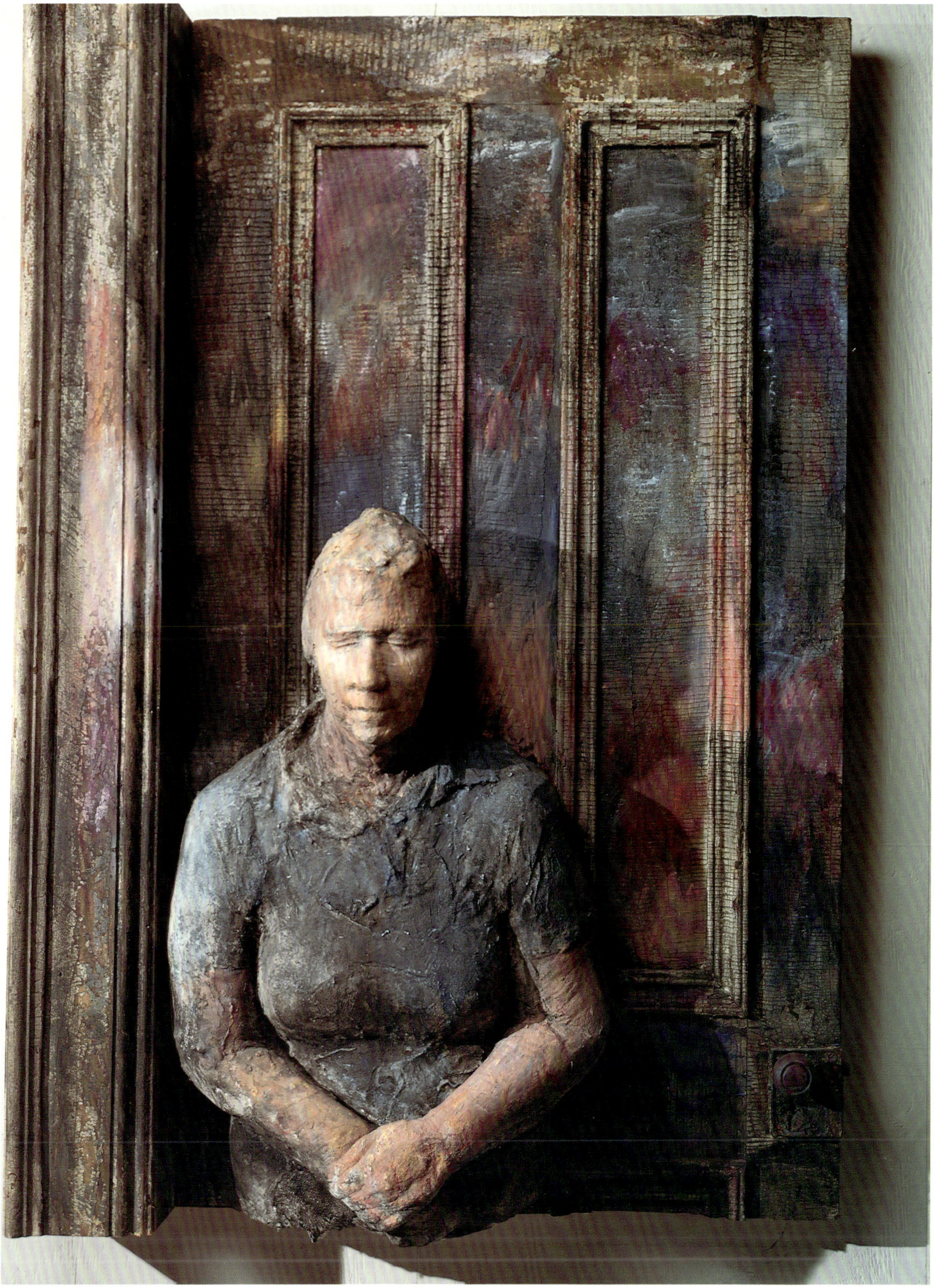

GEORGE SEGAL
Woman Against Multi-Colored Doorway, 1989
Plaster, wood, paint, metal
55 x 39 x 11 inches
Courtesy of Sidney Janis Gallery, New York

GEORGE SEGAL

1924 Born in New York City

Lives and works in New Brunswick, New Jersey

1970 Honorary Ph.D. Fine Arts, Rutgers University, New Jersey

1963 M.F.A. Rutgers University, New Jersey

SELECTED SOLO EXHIBITIONS

1989, 1988, 1986, 1984, 1982 1980, 1978, 1977, 1974, 1973, 1971, 1970, 1968, 1967, 1965 Sidney Janis Gallery, New York

1989 Modern Art Museum of Fort Worth, Texas; Orlando Art Museum, Florida; Portland Art Museum, Oregon

1983 The National Museum of Art, Osaka, Japan

1978-1979 "George Segal Retrospective," Walker Art Center, Minneapolis; San Francisco Museum of Modern Art; Whitney Museum of American Art, New York

1971-1973 European Museum Tour including the following museums: Kunsthaus Zurich; Musée des Beaux-Arts, Switzerland; Hessiches Landesmuseum, Darmstadt; Museum Boymans-van Beuningen, Rotterdam; Stadtisches Museum, Leverkusen, Germany; Centre National d'Art Contemporain, Paris; Stadtische Galerie, Munich; Palais des Beaux-Arts, Brussels

1959, 1958, 1957, 1956 Hansa Gallery, New York

SELECTED GROUP EXHIBITIONS

1983 "Controversial Public Art: From Rodin to di Suvero," Milwaukee Art Museum, Wisconsin; San Francisco Arts Commission, California

1975 "Sculpture: American Directions 1945-75," National Collection Of Fine Arts, Washington, D.C.

1968 "Dokumenta IV," Kassel, West Germany

1967 "American Sculpture of the Sixties," The Los Angeles County Museum of Art; Philadelphia Museum of Art

1963 "American Pop Art," travelling exhibition: The Hague; Ghent; Stedelijk Museum, Amsterdam; Moderna Museet, Stockholm

1962 "New Realists," Sidney Janis Gallery, New York

In all of his sculptures and tableaux, George Segal interprets a basic question posed by Cubism: what is real and what is illusion? By contrasting white plaster figures, color, real objects, and architectural fragments, the artist expands the conceptual and formal aspects of Picasso and Braque's art. In his environments, Segal invites the viewer to take their place, compelling us to distinguish between the realities of the plaster figure and our own existence. In this way, Segal establishes an important expressive as well as formal, compositional dynamic.

Although identified with the casting of live models, Segal modelled his earliest figures in plaster supported with chicken wire armatures. These are among his rawest and most expressive sculptures; they contain the forms and ideas from which much of his later work evolved. Working in plaster, with its references to tradition and figurative art, helped free Segal from the dominating influence of Abstract Expressionism. Although the artist rejected the movement's formal language, he accepted its philosophical premise: making art is a struggle which leads to the discovery and definition of self. *Man in Elevator* (1958-59) suggests the constriction felt by younger artists during this time. The figure, mummified, confined and fenced off by old doors, is marked by discrete globs of paint from the artist's palette. They do not run down the figure with Abstract Expressionist panache, but remain firmly coagulated. *Man in Elevator* is a humorous image, but it is also charged with a poignant tension.

Throughout his career, Segal has struggled to unify painting and sculpture in works where the monumental and the intimate expressively coexist. Beginning with *The Legend of Lot* (1958) and continuing with his most recent relief, *Woman Against Multi-colored Doorway* (1989, page 17), the artist weaves these aspects together through personal interpretations of art history. In the process of painting *Woman Against a Multi-colored Doorway*, Segal continues his meditations on Cubism by once again engaging perceptions of reality and illusion. The figure, a plaster cast of a living model, is interpreted conceptually. The painting on the figure and the choice of colors counter the apparent naturalism of the image. By contrast, the door, a real object, is treated more like a canvas. The artist paints this element more broadly, clearly articulating each brushstroke. The interaction between these two approaches activates the surface. Segal surrounds his peaceful and meditative portrait of a woman with a field of vibrating color. Her containment becomes more pronounced when contrasted with the artist's dynamic gestures. The problems of integrating painting and plaster sculpture are resolved.

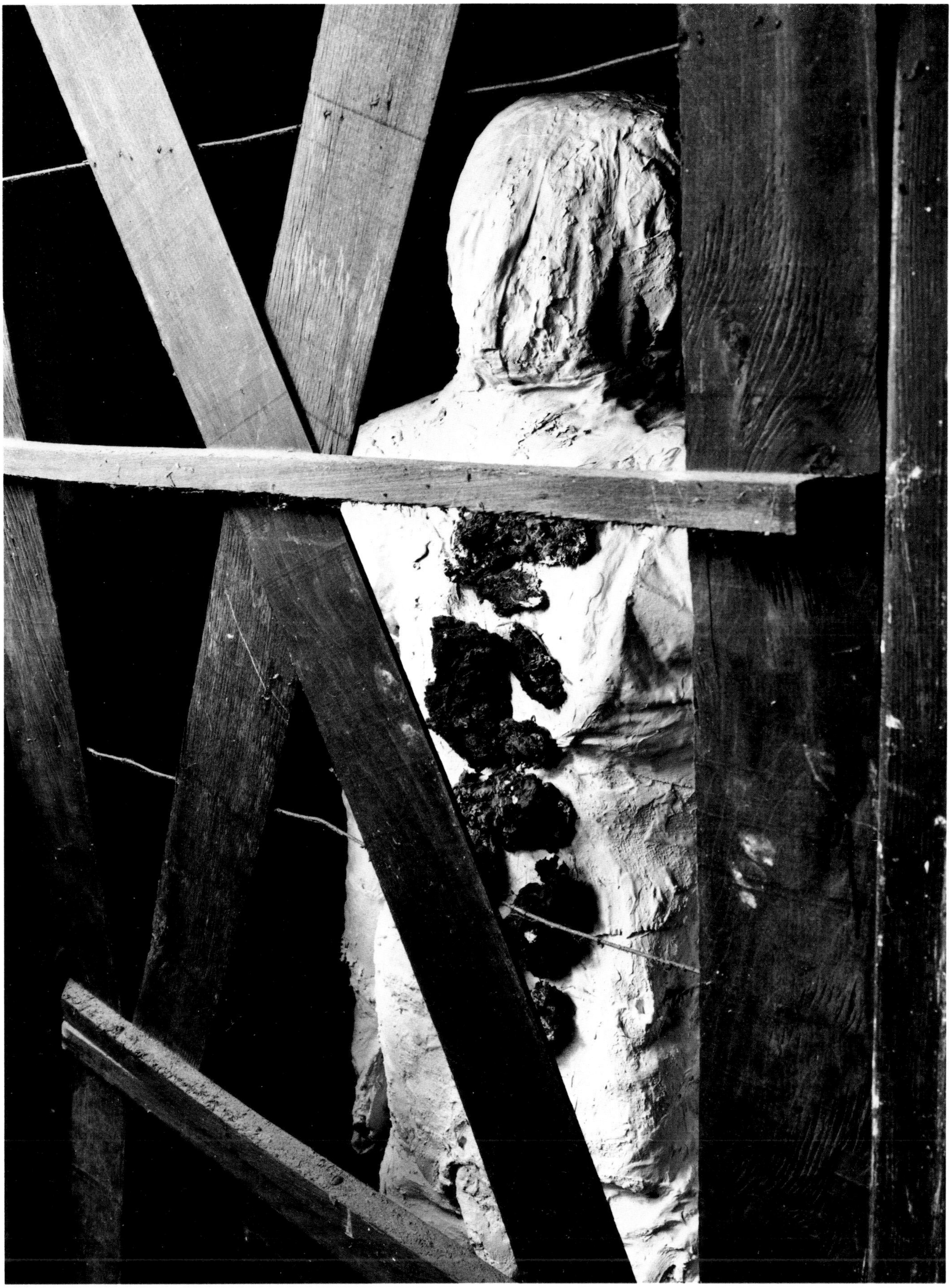

GEORGE SEGAL
Man in Elevator, 1958-59
Plaster on wire armature, wood, glass, paint
83 1/2 x 30 x 33 1/2 inches
Courtesy of Sidney Janis Gallery, New York

1956 Born in Baltimore, Maryland

Lives and works in Old Chatham, and Brooklyn, New York

1980 M.F.A. Columbia University Graduate School of Painting and Sculpture, New York

1974-78 B.F.A. Pratt Institute, Brooklyn, New York

SELECTED AWARDS

1987 First Alternate Award, Prix de Rome, The American Academy in Rome.

1986 First Prize Annual Art Competition, Loyola College, Baltimore, Maryland

1978 First Place: Drawing, juried exhibition, Pratt Institute, Brooklyn, New York

1977 Ford Foundation Fellowship

SELECTED SOLO EXHIBITIONS

1990 "Fragments and Pieces," Sculpture & Drawings, Hollywood Art & Culture Center, Florida

1990 "Drawings & Sculpture," New York Studio School,

1989 "Small Sculptures," Gaumann Cicchino Gallery, Ft. Lauderdale, Florida

1988 "Sculpture," 14 Sculptors Gallery, New York

1983 "Sculpture," 55 Mercer Street Gallery, New York

1982 "Sculpture & Drawing," Usdan Gallery, Bennington, Vermont

SELECTED GROUP EXHIBITIONS

1989 Group Drawing Exhibition, New York Studio School

1989 Faculty Drawing Exhibition, Parsons School of Art & Design, New York

1988 Commission for Orin & Elizabeth Raphael Sculpture Park, Chautauqua, New York

1984 "Architectural Sculpture," Germans Van Eck Gallery, New York

ED SMITH

Ed Smith believes in the enduring power of ancient art and myth to help define human identity and individuality. By reinterpreting classical figures from the Parthenon through abstract wood and plaster sculptures, the artist effectively synthesizes a traditional, heroic image with the realities of contemporary life. *Warrior Metope* (1987), removed from its original communal grouping, appears majestic, isolated and fragmentary. It reflects the struggle to order life's irrationality, the force that motivated ancient Greek artists and philosophers.

Smith interprets the continuing quest to transcend chaos through his process and forms. In *Warrior Metope*, the oozing, unformed plaster that the artist calls "mud," overflows the boundaries of the wood support. Smith contrasts the indeterminacy of cascading and recycled fragments of plaster with the stability of building joists. By simultaneously depicting movement and balance, Smith abstractly refers to the concept of classical *contrapposto*, an important innovation introduced into sculpture of the fifth century B.C. As a result, the artist injects a dynamic sense of equilibrium into his work.

Smith resolves his personal and physical struggle to integrate content, form and structure by a process of "throwing everything in and then paring it down." Consequently, *Warrior Metope*, like most of the artist's sculptures, is both constructed and modelled. Smith builds a frame and applies the plaster thickly, sometimes randomly, with a trowel and by hand. The wire lathe to which the material is applied remains visible as a raw, compositional element. The initial chaos of combining different materials and methods is refined through an intuitive sense of rhythm, pattern and balance as the artists seek harmony among the various elements.

The artist studied painting at Columbia University before working as assistant to sculptor, William Tucker. The influence of his early training is evident in his painterly approach to sculpture. Smith often mixes dry pigment into the wet plaster. In *Warrior Metope*, the artist also paints the dried surface in blue, pink and beige pastel tones. These colors emphasize the ethereal and airy aspects of the plaster, setting off the heavy, rustic wood structure. By painting the plaster, Smith introduces a poetic counterpoint to his sculpture's dominating presence.

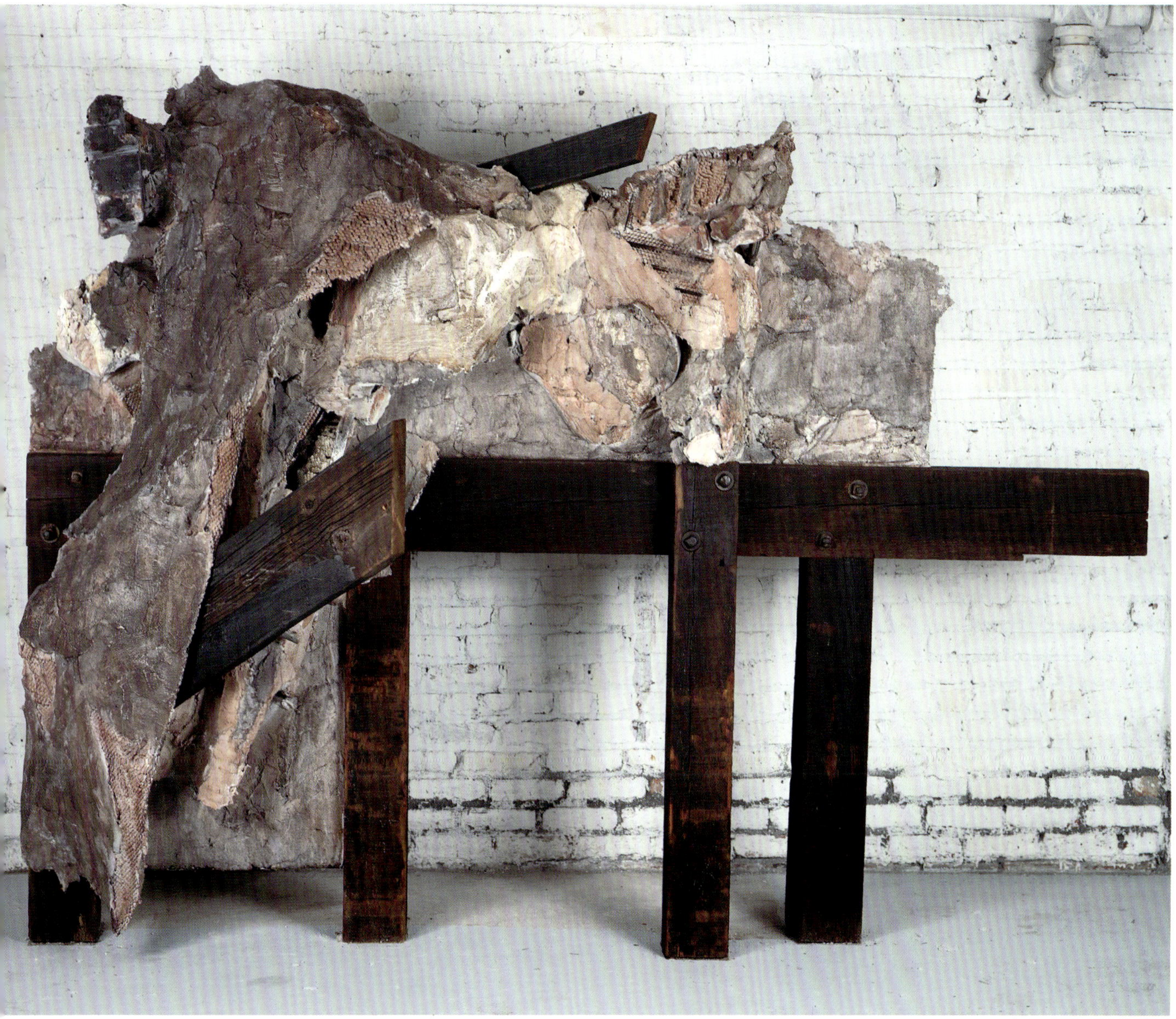

ED SMITH
Warrior Metope, 1987
84 x 108 x 48 inches
Structolite, steel, pigment

CLAUDE FIXLER

1954 Born in New York City

Lives and works in Waterville, Ohio

1981 M.F.A. Sculpture, University of Georgia, Athens

1976 B.A. Psychology, Washington & Lee University, Lexington, Virginia

SELECTED AWARDS

1986 National Endowment for the Arts/Arts Midwest

1986 Ohio Arts Council

1985 Greater Columbus Arts Council

1985 Columbus Museum of Art

SELECTED SOLO EXHIBITIONS

1988-89 Ohio State University Gallery, Newark, Ohio

1988 Bowling Green State University, Ohio

1986 Columbus Museum of Art, Beatty Corrider, Columbus, Ohio

SELECTED GROUP EXHIBITIONS

1988 "Choices: Twenty Painters from the Midwest," Evanston Art Center, Illinois; Wustum Museum of Fine Arts, Racine, Wisconsin

1988, 1987, 1986, 1985 Heritage Village Outdoor Sculpture Exhibition, Heritage Village, Columbus, Ohio

1987 "Third Annual Governor's Residence Art Collection," Columbus, Ohio

1985 "Columbus Art League Show," juried by Lowery Sims, Columbus Museum of Art, Ohio

1985 "Five Sides and Some Tops," Southern Ohio Museum, Portsmouth

1985 "Artists' Toys," Cleveland Center for Contemporary Art, Ohio

1984 "First Wyoming Biennial," juried by Dore Ashton, travelling throughout the state of Wyoming

1982 Alexandria Sculpture Festival, City of Alexandria, Virginia

SELECTED COMMISSIONS

1986 Ohio Historical Society, Columbus

1983 1800 M Street Medical Building, Washington, D.C.

Claude Fixler penetrates and reveals the dark fringes of the human mind. His androgynous plaster figures enact our recessed thoughts and feelings. Site-specific placements, inside or outdoors, intensify the psychological effect. For his Queens Museum installation, *Putting Ourselves to Bed* (1990), Fixler strategically exploits the dimensions of a small, narrow room to accentuate a sense of confinement. He compels the viewer to witness a ritual of burial and death which symbolizes our memories and our past.

We pass through a blackened corridor to the intimacy of a private sanctuary. Along the two long sides of the rectangular space, life-size plaster figures, white and unearthly, lean in succession. A ramp leads down to the far wall where rows of additional figures recline in their individual niches, stacked, as in a catacomb. A mound of plaster is beginning to inundate the space, covering bodies, and consigning them to oblivion. The effect is eerily surreal. Unexpectedly confronting this ceremony of the souls makes us ask: What is our place here? We are strangers to the scene, distanced by the otherworldly whiteness of the characters; nevertheless, we are gripped by the common chords of fear and fascination that surface in any encounter with human mortality.

In all of Fixler's plaster installations, the public participates in a ritual of discovery. The interaction of people, plaster figures, space and architecture is crucial, not only to the composition, but also to its meaning. Through this dynamic interchange, the viewers take their place among the white apparitions which mirror our inner selves.

This effect is created by the artist who models his figures in clay, then casts them in plaster. The whiteness of the plaster imparts a disturbing view of the human form. The figures appear immobilized and alienated, yet at the same time, their connection to humanity is implicit. The shell-like fragility of the thin plaster cast exists on the brink of disintegration, maintaining a tenuous and poignant link to our own world.

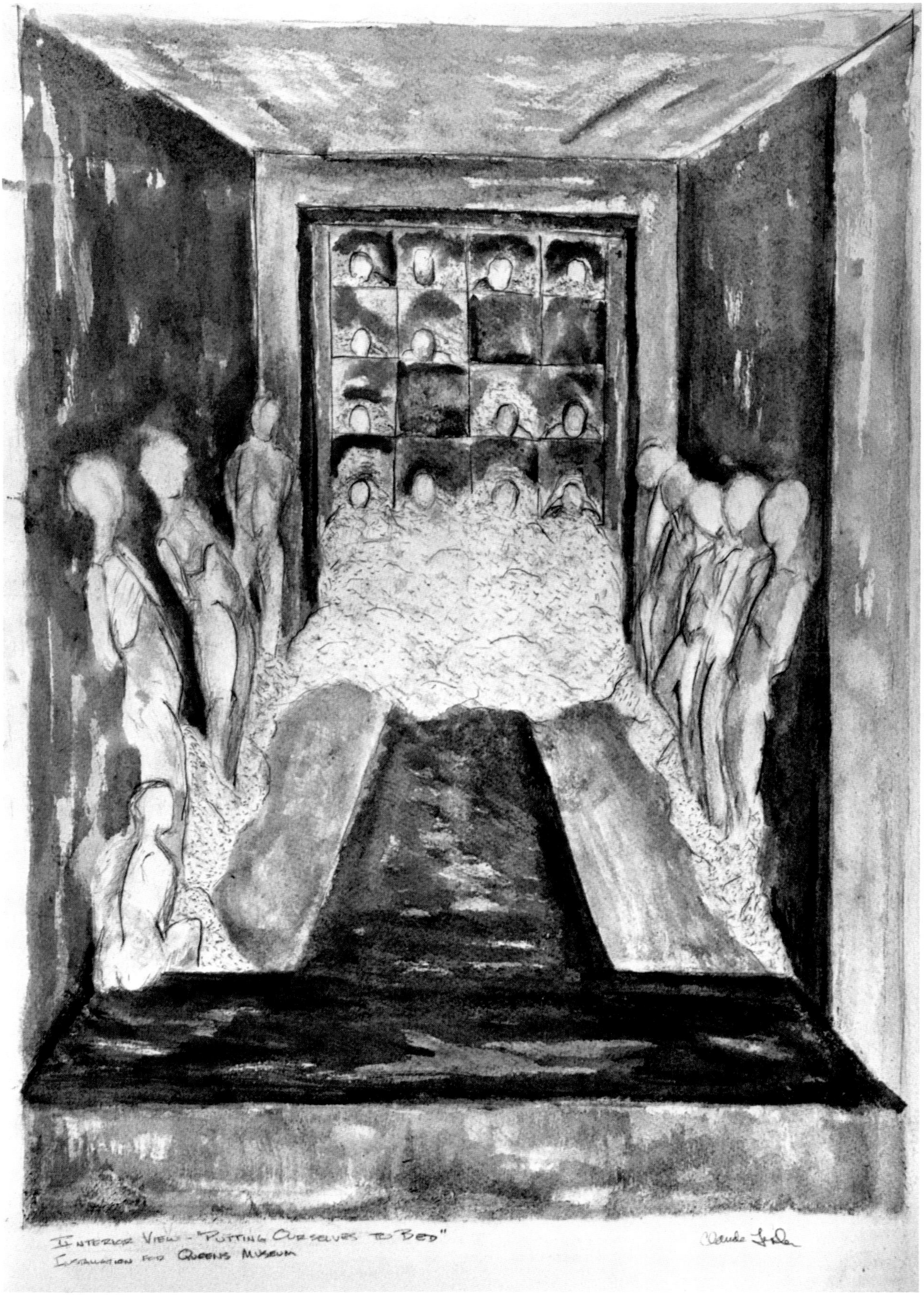

CLAUDE FIXLER
Drawing for: *Putting Ourselves to Bed*, 1989-90
Site-specific installation for the West Gallery
Hydrocal, plaster gauze, wood
Drawing: Ink wash and graphite on paper
22 x 15 inches

ANITA CURTIS GLESTA

1958 Born in Chicago, Illinois

Lives and works in Brooklyn, New York

1980-82 Boston Museum of Fine Arts School

1977 Skowhegan School of Painting and Sculpture

1976-79 B.A. Literature & Philosophy, Seminar College at The New School for Social Research

1974-75 Kirkland/Hamilton College, Clinton, New York

SELECTED SOLO EXHIBITIONS

1989 Hudson River Museum, Yonkers, New York

1985 White Columns Gallery: The White Room, New York

1985 Johansson Gallery, Uppsala, Sweden

SELECTED GROUP EXHIBITIONS

1990 "Forces of Nature: Outdoor Sculpture," Long Island University, Brooklyn, New York

1990 "Sculpture Installation," Organization of Independent Artists, The Federal Building, New York

1987 "Painting into Sculpture," curated by Charlotta Kotik, The Brooklyn Museum at The Rotunda Gallery, Brooklyn, New York.

1987 "Neo-Alchemy: Works By Visionary Artists," Cavin Morris Gallery, New York

1986 "Painting in the Third Dimension," City Without Walls Gallery, Newark, New Jersey

1985 "Gallery Artists," White Columns Gallery, New York

1985 "Shifting Views," BACA Gallery, Brooklyn, New York

1983 "The Terminal Show," curated by Ted Castle, Brooklyn, New York

Anita Curtis Glesta's abstract plaster sculptures conjure primal, archetypal memories. Swathed and cradled in cheesecloth, these textured embryos balance or hang suspended from wood and painted canvas. Glesta uses these shapes as formal and psychological foils to the rigidity of the other materials in her composition.

Since 1985, Glesta has been working with plaster to redefine the charged emotions generated by Western and Eastern mythology. Moving from references to specific heroes, goddesses and legends, she now visualizes more universal, organic metaphors. To elicit a gut reaction, the artist creates tactile, visceral sculptures in plaster. Glesta describes the emotional impact of her work by analogy with a swamp. Initially, we are repelled by what appears to be an ugly, messy and forbidding environment. Once acclimated, the full spectrum of life is subtly revealed to us.

Like many artists in this exhibition, Glesta began her career as a painter. From thickly painted canvases, to reliefs, and then actual three-dimensional objects projecting from the canvas, she evolved a more sculptural language to complement the content of her work. Her compositions grew more dynamic and the stream of associations deepened.

Plaster catalyzed Glesta's integration of painting and sculpture. Although plaster constructions and brightly painted canvas retain their separate identities, pigmentation of the plaster unifies the two. While the plaster is semi-wet, the artist rubs acrylic paint into the material. This technique and her choice of earth tones suggest a state of fossilization. Evocations of an ancient past resonate through color and form.

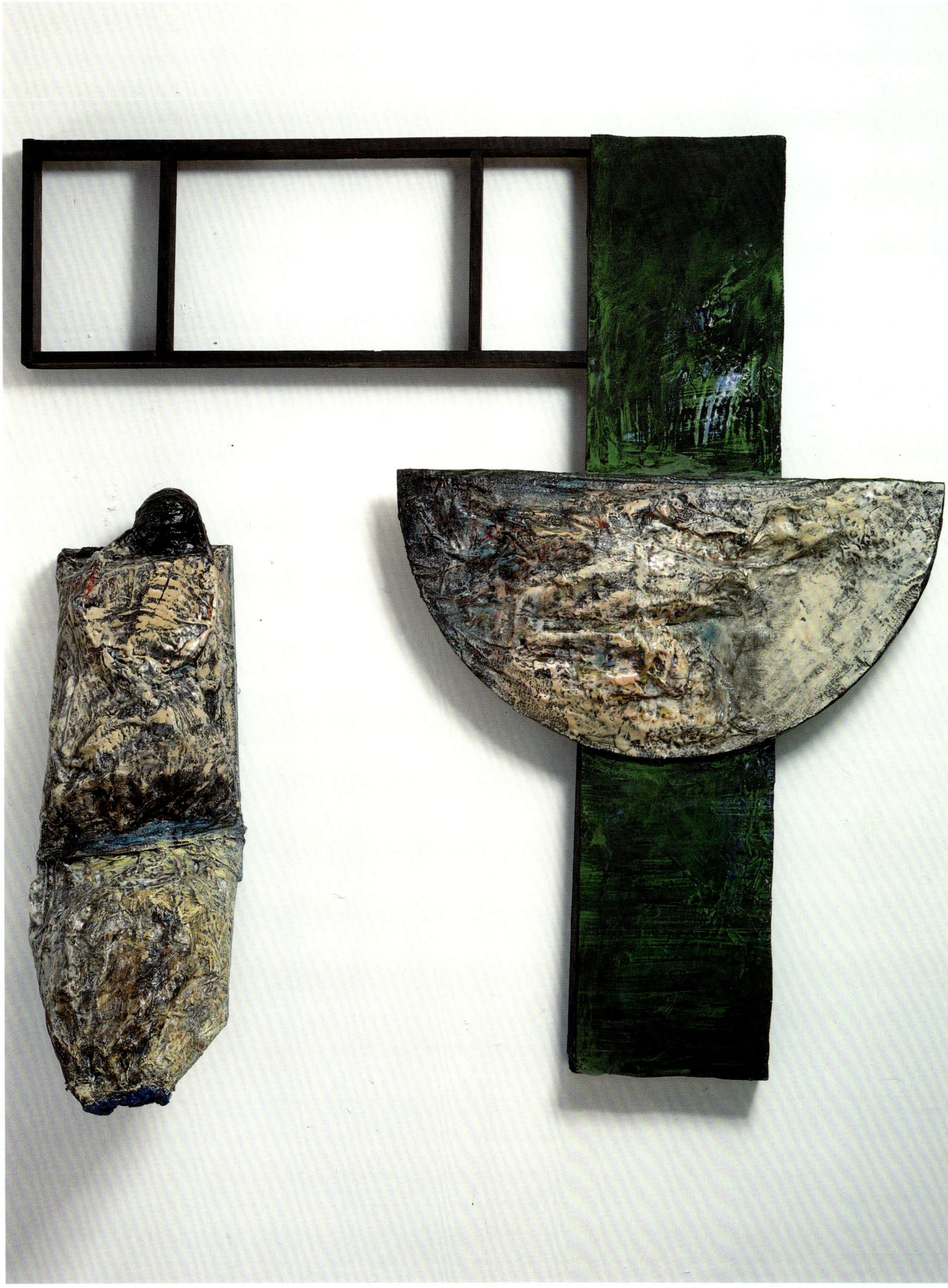

ANITA CURTIS GLESTA
Seated Figure, 1989
Plaster and mixed media
70 x 56 inches

BARRY GOLDBERG

1959 Born in New Brunswick, New Jersey

Lives and works in New York City

1981 B.A. Cooper Union, New York

SELECTED AWARDS

1989, 1987 Pollock-Krasner Foundation Grant

1981 The Ethel Cram Memorial Prize in Art, Cooper Union

SELECTED SOLO EXHIBITIONS

1989 55 Mercer Gallery, New York

1985 "Special Projects," P.S. 1, Long Island City, New York

SELECTED GROUP EXHIBITIONS

1989 Arnold Herstand & Co., New York

1987, 1986 Sculpture Center, New York

1985, 1984 Paulo Salvador Gallery, New York

1985 Lidewij Edelkoort Gallery, Paris

In Barry Goldberg's sculptures, paradoxical readings of movement and stasis, destruction and meditation, collide. The artist intimates these connections through an interplay of formal relationships. This is apparent in *Drifting Buoy* (1989) which marks location, balanced on a hybrid base of waves and wheels. A real hammer hangs ominously from a chain. With this tool, the artist has punctured the flat, upper section to simulate stars in the night sky, a reference to navigation and wish-making. By incorporating the hammer, Goldberg invites the viewer to participate in what he describes as a "fetishistic act; an uncertain mixing of violence and creation."

In *Heartwood* (1987), the artist conceptualizes the life process of a tree; it buries itself in concentric layers while reaching skyward to create a foliage display. This contradiction interests Goldberg who creates the tree's center, or heartwood, by suspending it from tentacled limbs. Like sap, it appears to drain into a vessel, a symbolic offering of growth and regeneration.

Barry Goldberg's carefully planned and constructed sculptures evolve from his annotated notebook drawings. In these drawings, the multi-layered associations and meanings embedded in the final work find initial expression. Memories and events are poetically translated, with the help of cardboard and clay models, into tall plaster sculptures. In most of his works, Goldberg constructs an armature of frontally oriented, vertical steel members that allude to figural and architectural features. Once the armature is assembled, the artist works extensively with jute dipped in pigmented plaster. The finished surface, modulated by variations in drying, includes ink, paint, and additional plaster to heighten the tonal range and expressive content. Goldberg channels plaster's flexibility to create fine, precise, tensile lines. His spiky, curvilinear forms contrast with the volume and mass normally associated with the free flowing qualities of the material.

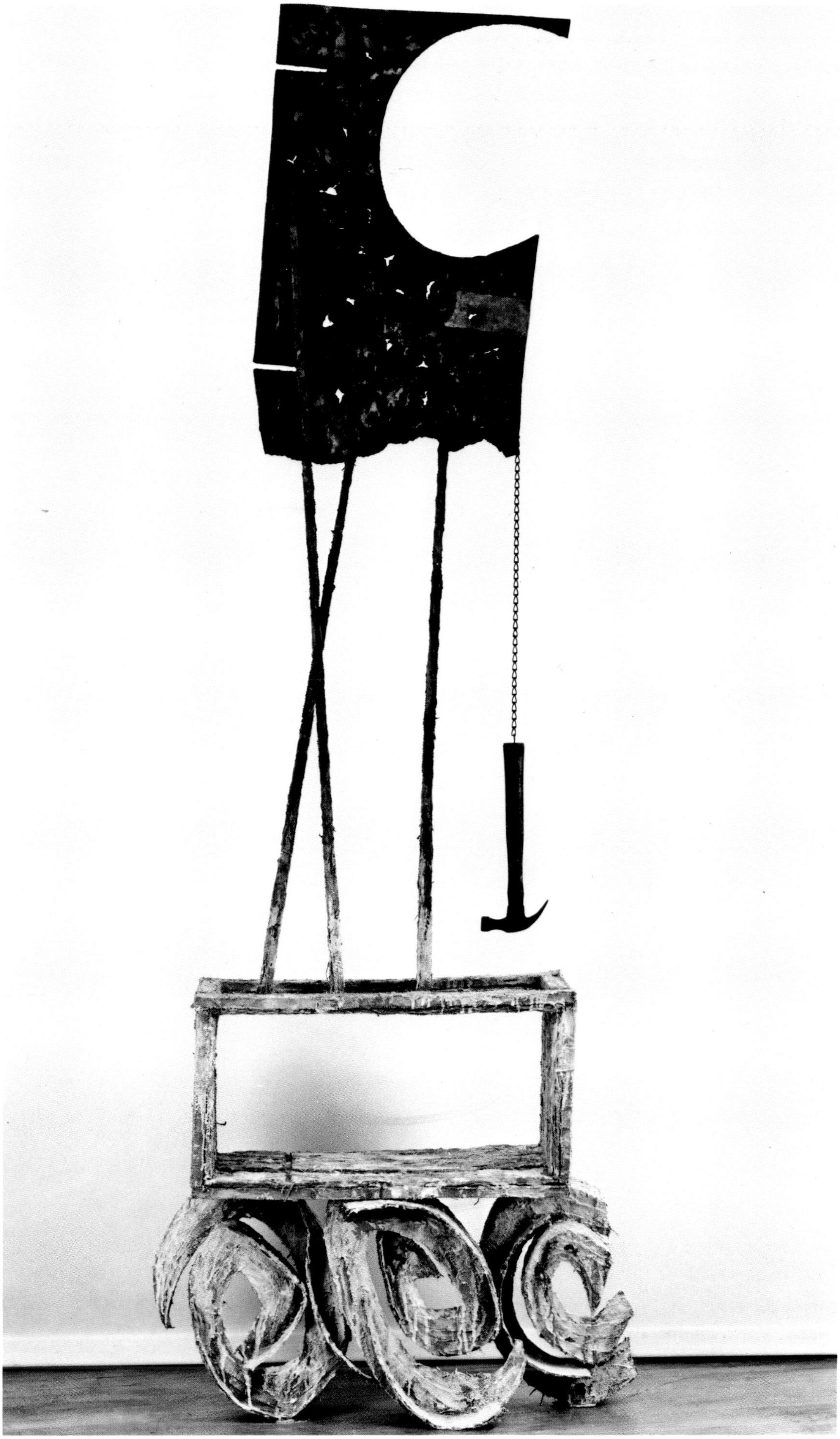

BARRY GOLDBERG
Drifting Buoy, 1989
Hydrocal, steel and mixed media
100 x 36 x 12 inches

GRACE KNOWLTON

1932 Born in Buffalo, New York

Lives and works in Snedens Landing, Palisades, New York

1981 M.A. Art & Education, Columbia University Teachers College, New York

1954 M.A., B.A. Art, Smith College, Northampton, Massachusetts

SELECTED AWARDS

1987 Public Art Fund, Inc.

SELECTED SOLO EXHIBITIONS

1989 East Hampton Center for Contemporary Art, Long Island, New York

1989 Bill Bace Gallery, New York

1987 City Hall Park, New York

1987-1988 Hillwood Art Gallery, Long Island University, Brookville, New York

1986, 1984 Twining Gallery, New York

1979 Parsons-Dreyfuss Gallery, New York

1977 Elaine Benson Gallery, Bridgehampton, New York

1973 Henri Gallery, Washington, D.C.

1971 Spectrum Gallery, New York

SELECTED GROUP EXHIBITIONS

1989 Blue Hill Cultural Center, Pearl River, New York

1988 Twining Gallery, New York

1987 Socrates Sculpture Park, Long Island City, New York

1986 Aldrich Museum of Contemporary Art, Ridgefield, Connecticut

1982 Roberson Center for the Arts, Binghampton, New York

1980 Organization of Independent Artists, Wards Island, New York

1978 Rutgers University, Camden, New Jersey

1972 Spectrum Gallery, New York

1969 Henri Gallery, Washington, D.C.

Grace Knowlton's *Sphere* (1989) and *Shard* (1988), exhibited at the entrance to The Queens Museum in Flushing Meadows Park, harmonize with landscape settings — places where humanity, nature and history intersect. They can be read as geographical markers, like boulders, or the ritualistic remains of early inhabitants. Never perfectly round, Knowlton's sculptures bulge and contract along their textured surfaces, often a field for her boldly abstract painting. For The Queens Museum, Knowlton wanted "the message to be plaster," so she left the *Sphere* white, and exposed the layers of its construction — lathe, wire mesh, burlap, and fiberglass slathered with Design Cast plaster.

Knowlton began her career as a painter after studying with Kenneth Noland in Washington D.C. during the 1960s. Her canvases became more three-dimensional when latex rubber forms cast from clay pots were attached to the surface. Interested in the concept of enclosure, Knowlton embarked on her series of spheres. These were made out of a variety of materials: terra-cotta, Corten steel and copper. Four years ago, the artist discovered Design Cast, a variety of plaster durable enough to exhibit outdoors.

In a new series entitled *Tracings* (1989), Knowlton started casting sections of glacially scoured granite outcrops located on her country property. The natural history of the rock is recorded by the artist whose own layering process, a method which includes intentional breaks and restorations in the plaster, parallels the subject's geological origins.

Knowlton's equation of man, woman and nature continues to evolve in *Triptych* (1989), three vertical plaster panels that lean against the wall. This work is more conceptual, less literally bound to natural forms, and Knowlton cast it from styrofoam rather than rock. Still, the materials mixed into and attached to the plaster — mud, clay, sticks, grass, and iron pigment — refer to growth and decay. The plaster panels, honeycombed and randomly speckled with stray styrofoam pellets, are light and ethereal. As in most of Knowlton's sculptures, surfaces are delicately thin, a membrane filtering meditations on time and the cycles of nature.

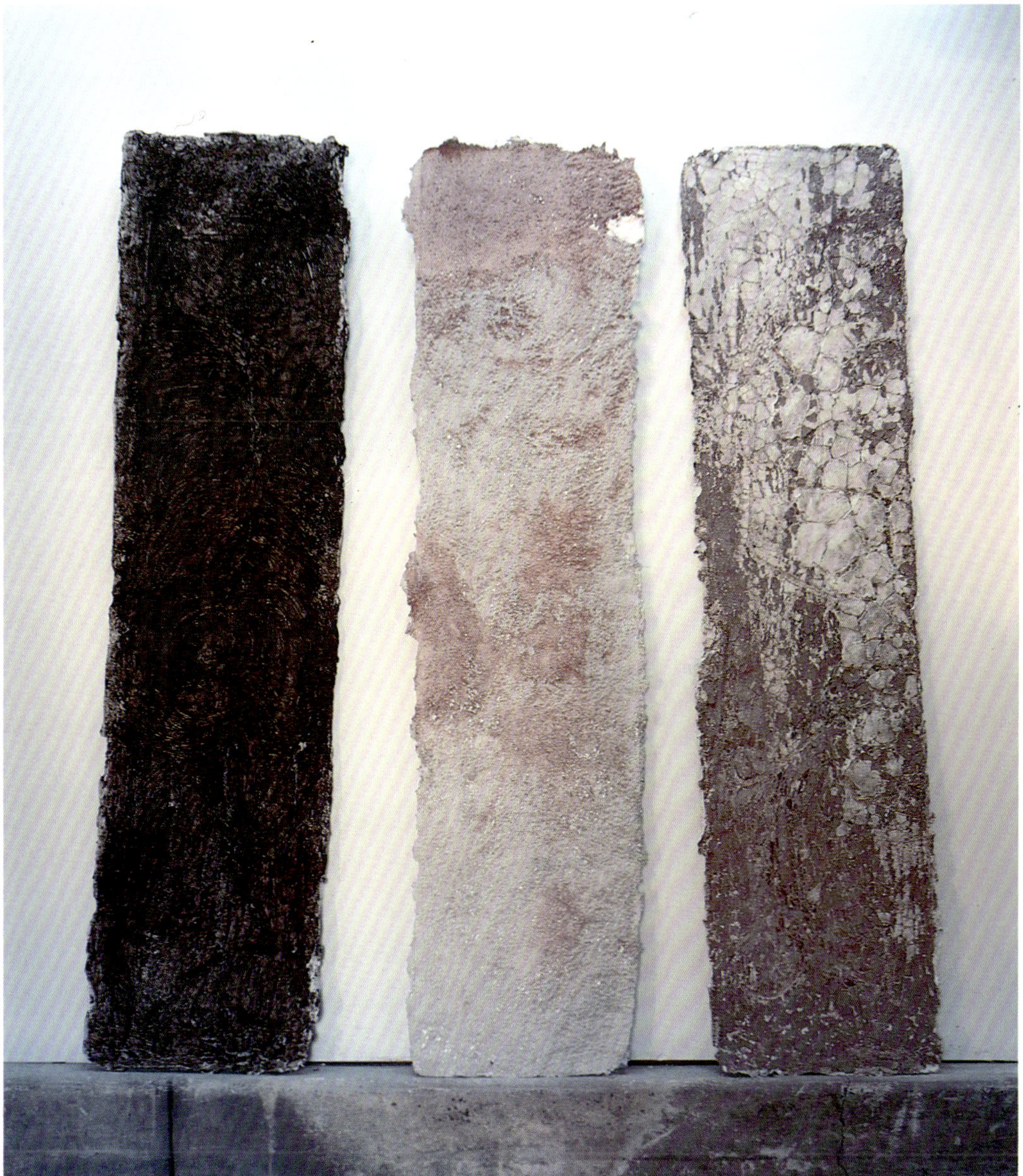

GRACE KNOWLTON
Triptych, 1988
Design Cast, mud, iron, pigment
Each panel: 96 x 24 x 1/2 inches

RUDOLPH SERRA

1948 Born in San Francisco, California

Lives and works in New York City

Teaches at Bennington College, Vermont

1975 M.F.A. University of California, Berkeley

1973 M.A. University of California, Berkeley

1971 B.A. San Francisco State University

SELECTED AWARDS

1984, 1978, 1976 National Endowment for the Arts Grant (Sculpture)

SELECTED SOLO EXHIBITIONS

1990 John Davis Gallery, New York

1987 Sculpture Installation, Kenmare Square, New York

1985 Outdoor Sculpture, The Nassau County Museum of Fine Art, Roslyn, New York

1982 Baruch College, New York

1980 The University of Houston, Texas

SELECTED GROUP EXHIBITIONS

1989 "Group Exhibition," Tomoko Liguori Gallery, New York

1988 "Cube Exhibition," University of Massachusetts at Amherst

1987 Socrates Sculpture Park, Long Island City, New York

1983 The 12th International Sculpture Conference, Oakland Museum, Oakland, California

1979 "Mark Di Suvero, Rudolph Serra, Bill Wareham," Hansen Fuller Gallery, San Francisco

1977 "Eighteen Bay Area Artists," University Art Museum, Berkeley, California

1975 "Whitney Biennial," Whitney Museum of American Art, New York

1974 "Two Person Exhibition: Rudolph Serra and Deborah Butterfield," University Art Museum, Berkeley, California

Rudolph Serra's sculptures are abstract monuments that stake out places of refuge and meditation. Many works, such as *Momraw Series #2: Cube* (1988), link wall and floor through an unusual perspective effect. The tumultuous exterior surface drips downwards and sideways to the squared opening at the sculpture's base. Glancing into this cavity, the viewer experiences the mystery and beauty of shifting geometries. Visible access to the dark, recessed spaces defined by plaster is fundamental to both the meaning and composition of the work.

Serra's work in plaster directly reveals the energy and physicality of his process. When his mixture reaches the critical consistency, he throws it on the sculpture, creating an agglomerated surface not unlike the stalactites and stalagmites of cave formations. Many effects are left to chance, as gravity determines the final texture. The artist admires the works created by his older brother, Richard Serra, whose *Splash* pieces (1968) epitomized the radical methods of Process Art. Whereas Richard threw molten lead into the space between wall and floor with few preconceptions about the final results, Rudolph throws plaster on an abstract sculpture already constructed. For the younger Serra, the flow of plaster represents freedom; working with it is a way to break out of rigid forms and thoughts.

While Serra's process is intuitive and often governed by accident, the underlying forms into which his plaster is poured are carefully planned and painstakingly constructed. The artist designs small models of cardboard and string which he translates into elliptic and geometric wood shapes. These are removed after a cast is made, and the process of throwing and sanding the plaster begins. Serra smooths and polishes the interior surfaces while leaving the external ones raw and active. Light and shadow accentuate the elegant, sharp-edged boundaries and the surrounding web of textured mass. Although Serra has been working with plaster since 1977, his fusion of "cool," geometric structures with chaotically energized surfaces, reflecting dualities of emotion, represents a new direction.

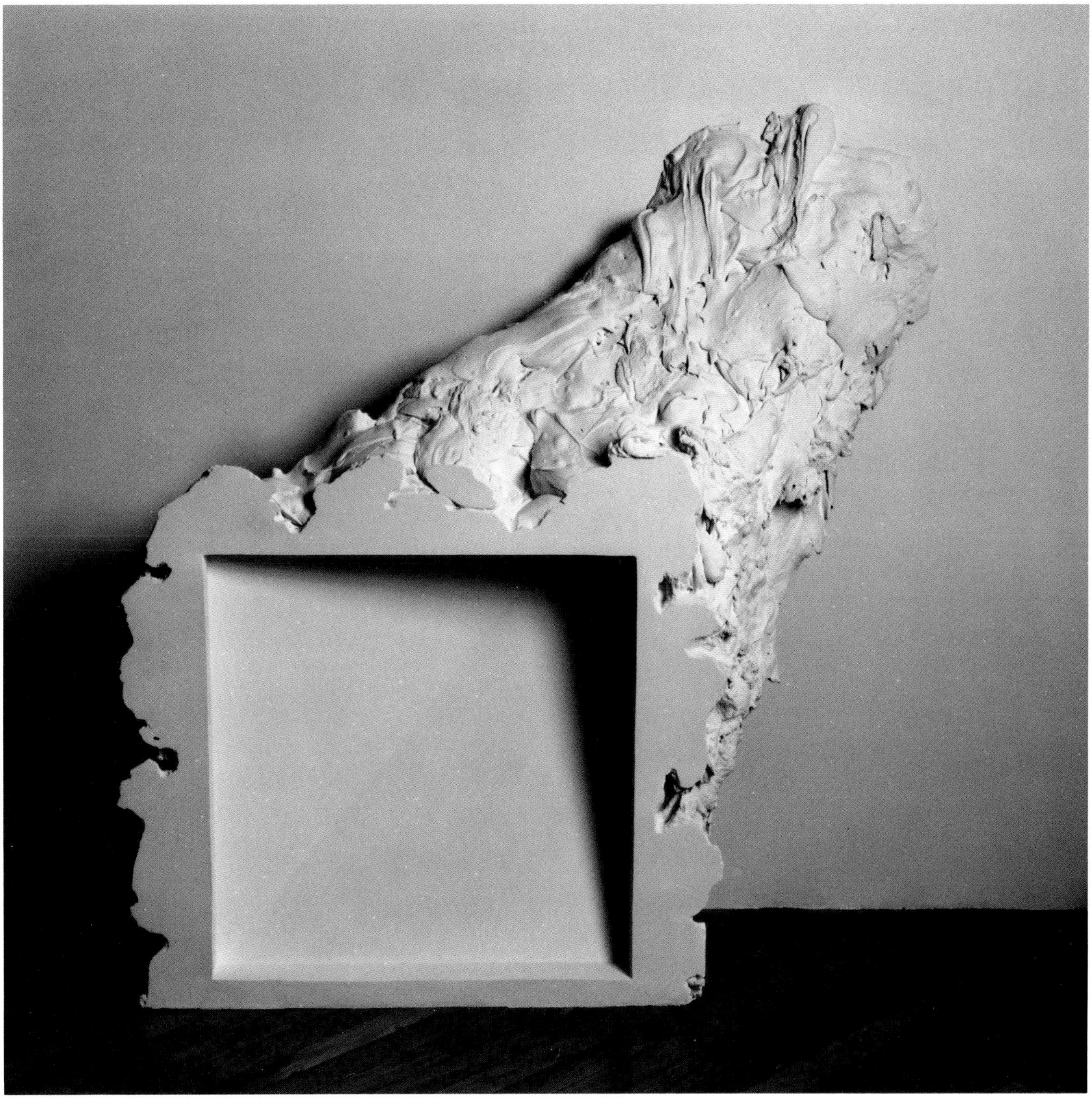

RUDOLPH SERRA
Momraw Series #2: Cube, 1988
Hydrocal
36 x 48 inches

SUSAN LAUFER

1950 Born in Tuckahoe, New York

Lives and works in New York City

1975 M.A. New York University

1972 B.F.A. Boston University

SELECTED AWARDS

1985 The Ferkauf Foundation, New York

1984 National Endowment for the Arts (Individual Artist Grant)

SELECTED SOLO EXHIBITIONS

1990, 1989, 1987, 1985, 1984 Germans Van Eck, New York

1989 Stephen Wirtz Gallery, San Francisco, California

1988 Barbara Krakow Gallery, Boston, Massachusetts

1988 John Stoller Gallery, Minneapolis, Minnesota

1986 vanStraaten Gallery, Chicago

SELECTED GROUP EXHIBITIONS

1990 "The Matter at Hand," University of Wisconsin Art Museum, Milwaukee

1990 "Working on Paper: Contemporary American Drawings," The High Museum, Atlanta, Georgia

1989 "One Hundred Women's Drawings," Hillwood Art Gallery, Brookville, New York and Blum Helman Warehouse, New York

1988 "New Painting/New York," curated by Kay Larson, Spencer Museum of Art, Lawrence, Kansas; Seattle Art Museum; University of California, Los Angeles

1988 "Lifeforce," Robeson Gallery, Rutgers University, Newark, New Jersey

1987 Group Exhibition, Albright-Knox Art Gallery, Buffalo, New York

1986 "Law and Order Show," Benefit Exhibition: Leo Castelli, Paula Cooper, Barbara Gladstone, John Weber Galleries, New York

1985 "Abstract Issues," curated by Steven Henry Madoff. Tibor de Nagy, Oscarsson Hood and Sherry French Galleries, New York

1984 "Contemporary Triptychs," Edith C. Blum Art Institute, Bard College, Annandale-on-Hudson, New York

1983 The Alternative Museum, New York

Upon entering Susan Laufer's studio, we are confronted with large-scale paintings in varying stages of completion, and a mound of mysterious artifacts which become her vocabulary of forms. These primeval shapes, cut from wood or found by the artist, are temporarily attached to masonite. Over the course of two years, the approximate time she takes to create a body of work, Laufer builds a surface of pigmented plaster around these elusive shapes which she later removes. The heavily encrusted templates are themselves evocative objects which the artist intends to cast in various metals. Their imprint materializes out of a swirling, densely textured atmosphere. While some areas of the painting are layered in relief, other sections are simultaneously eroded by sanding. Laufer creates undulating topographies which she scratches, carves, and draws upon. Her process and subject converge into a single drama of erupting and folding matter.

Laufer's imagery reflects her life-long interest in archaeology, paleontology and African sculpture, as well as her travels to Afghanistan, North Africa, Iran and, most recently, Egypt. A collection of bones, fossils, effigies and weathered machine parts line the desk and windowsills of her loft. The associative power of these objects are distilled in her paintings which can be interpreted as archetypal and cosmological landscapes.

Within Laufer's paintings, an internal light source appears to emit energy. Mica chips from the gypsum in plaster intensify this spectral glow. Laufer pigments the plaster in hues, ranging from warm ochres to intense ultramarine blues. Through color and texture, the artist defines a struggle between dark and light that mirrors an emotional and physical dialectic — turmoil and tranquility, expansion and containment. Since 1976 when Laufer began incorporating plaster into her paintings, she has developed a style that abstractly expresses the energies of macro- and microcosm by blurring distinctions of time, scale and space.

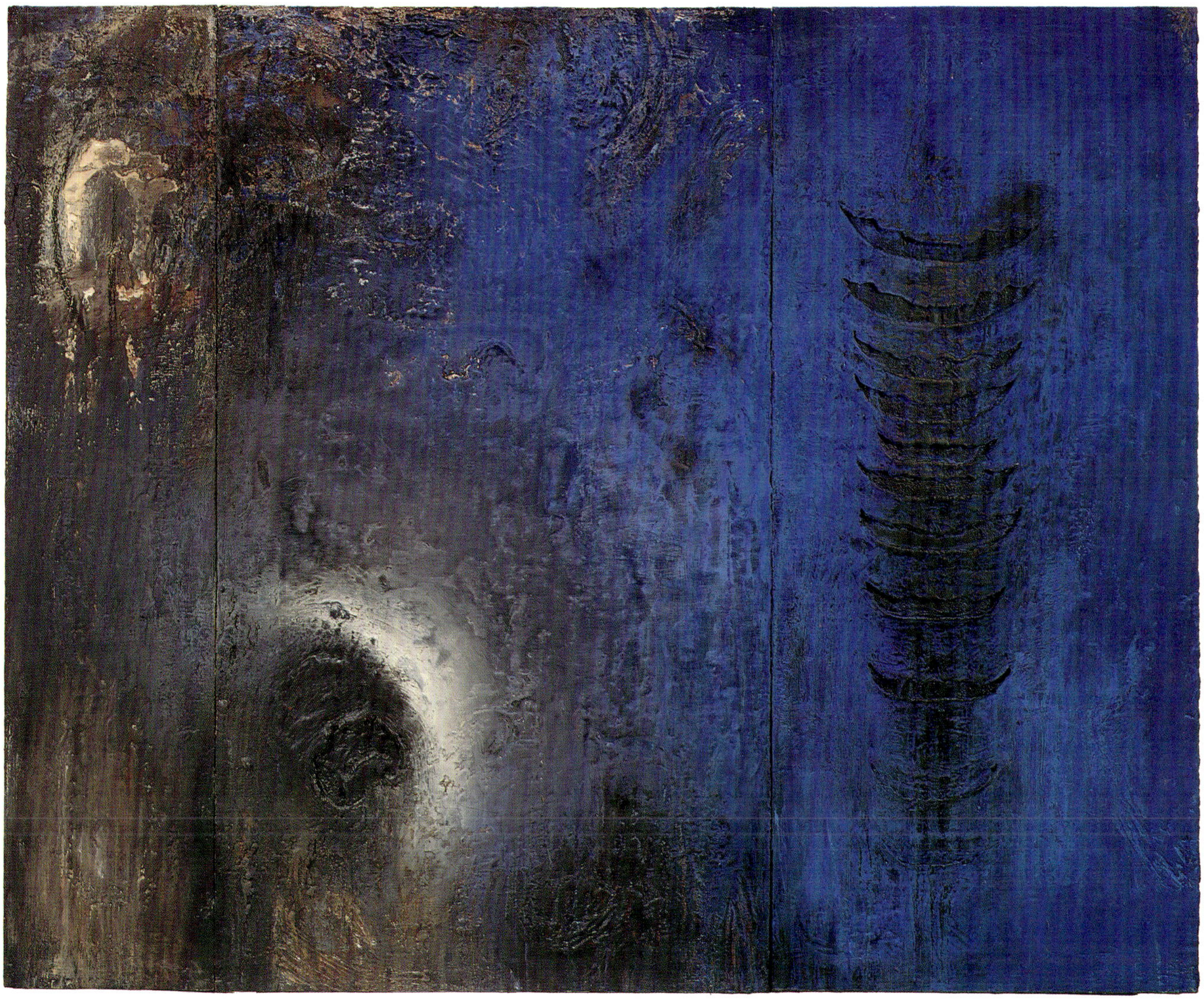

SUSAN LAUFER
Deep Blue, 1987-90
Acrylic, polymer, and plaster on panels
108 x 134 inches

JONATHAN SILVER

1937 Born in New York City

Lives and works in New York City

1968-1969 Edward J. Noble Fellow in Art, Columbia University

1966 M.A. Art History, Columbia University

1963 B.S. Columbia University

SELECTED SOLO EXHIBITIONS

1989 "The Lower Room: An Installation," Sculpture Center, New York

1989 Trinity Gallery, Atlanta, Georgia

1988 "Chance Overthrown," Montclair Art Museum, Montclair, New Jersey

1988 Gremillion Gallery, Houston, Texas

1987 University of Bridgeport, Connecticut

1987 Grimaldis Gallery, Baltimore, Maryland

1986 Gruenebaum Gallery, New York

1984 New York Studio School, New York

SELECTED GROUP EXHIBITIONS

1988 "American Baroque," Holly Solomon Gallery, New York

1987 "Romanticism and Classicism," Queensboro Community Gallery, Bayside, New York

1986 "Art On Paper Exhibition," Weatherspoon Art Gallery, University of North Carolina, Greensboro

1979 Two Person Exhibition with Christopher Cairns, New York Studio School, New York

1976 Two Person Exhibition with Christopher Cairns, 4x10 Gallery, New York

1976 Comfort Gallery, Haverford College, Haverford, Pennsylvania

SELECTED COMMISSIONS

"The Birth of Venus," Hopkins Plaza, Baltimore, Maryland

Jonathan Silver's work in plaster expresses a full spectrum of human emotions — love, violence, empathy and fear. For The Queens Museum, the artist will create an installation entitled *Subway*. Through macabre and ironical humor, he reveals a tableau of late twentieth-century urban life and the materials of his process. From black rubber molds, white plaster shells and ruddy clay mounds, amorphous, semi-human figures slither and ooze. They are witnesses, together with us, to the victimization of an aluminum-clad plaster figure as it is pushed from the subway platform. Tracks cast in plaster hang suspended from the ceiling as floor mirrors reflect the image into the viewer's realm.

In this modern tragedy, the artist secularizes the heroic, multi-figure monuments of the past. Silver's desire to restore sculpture to its most powerful position, in unity with architecture and narrative or symbolic context, developed when he studied Michelangelo's *Medici Chapel* (1520-34) and Bernini's *Cornaro Chapel* (1645-52) in Italy. Consequently, the artist no longer conceives sculpture as an independent object but as part of a dramatic ensemble.

In 1966, Silver began working in plaster with Peter Agostini at Columbia University while researching the paintings of Alberto Giacometti. The influence of Agostini, Giacometti and Rodin can be seen in the freedom with which the artist defines and activates his sculpture's surface and the surrounding space. Silver's works are often sensuously modelled and elongated, defined by a majestic sweep in the torso or limbs.

The artist models and combines clay and plaster, making plaster casts that are recycled and constructed into unique assemblages. This approach, shared by Christopher Cairns, expands Rodin's method of creating multi-figure compositions from sections of unrelated casts. In Silver's sculptures, abstract and representational aspects of the human form are unexpectedly juxtaposed. These enigmatic dislocations are both innovative and emotionally compelling. By so composing his individual figures then placing them in narrative environments, Silver revitalizes the tradition of the human figure in sculpture.

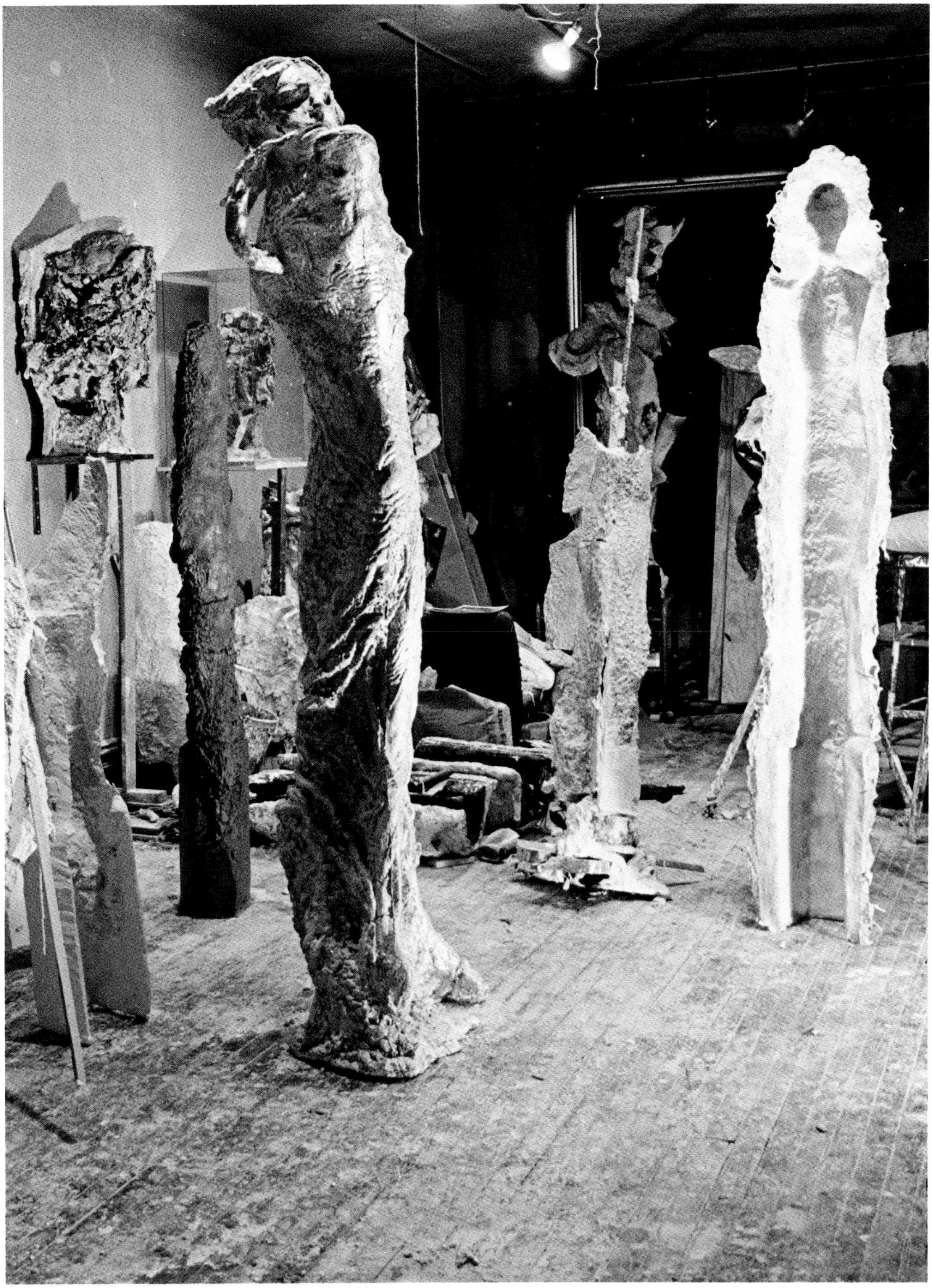

JONATHAN SILVER
View of artist's studio
In preparation for *Subway,* 1990
Site-specific installation for the West Gallery

ROBERT VAN VRANKEN

1960 Born in Sayville, New York

Lives and works in Richmond, Maine

1985 M.F.A. San Francisco Art Institute, California

1983 Bowdoin College, Brunswick, Maine

SELECTED AWARDS

1989 The Virginia Center for the Creative Arts, Artist's Residency

1988, 1986 The MacDowell Colony for the Arts, Artist's Fellowship

1988 The Millay Colony for the Arts, Artist's Residency

1987 The Elizabeth Greenshields Foundation, Artist's Grant

SELECTED SOLO EXHIBITIONS

1989, 1988 O.K. Harris, New York

1987 Bowdoin College Museum of Art, Brunswick, Maine

1985 Graystone Gallery, San Francisco, California

SELECTED GROUP EXHIBITIONS

1989 "Contemporary Masters," Glenn Dash Gallery, Los Angeles

1987 "ArtQuest 1987," Art Institute of Boston; California State University, Los Angeles

1986 "Summer Faculty Exhibition," Portland School of Art, Maine

1986 "Prints from the Permanent Collection," Bowdoin College Museum of Art, Brunswick, Maine

1986 "The North Dakota Annual Print and Drawing Exhibition," University of North Dakota, Grand Forks

1986 "The Boston Printmakers Annual Members' Exhibition," Duxbury Art Complex Museum, Massachusetts

1985 "The American Annual Works on Paper Exhibition," Zaner Gallery, Rochester, New York

1985 Graystone Gallery, San Francisco, California

1985 "Maine Artists," Wingspread Gallery, Northeast Harbor, Maine

1985 "Masters 85," Fort Mason Center, San Francisco, California

Robert Van Vranken builds his paintings like the old, disintegrating urban walls they often resemble. To achieve texture and dimension, he layers plywood with coats of plaster which are then gouged and incised. Irregularities resulting from the application of plaster energize the surface, together with stripes of dripped paint and collaged papers. Van Vranken uses plaster in a traditional way, as a ground for his painting. Unlike traditional artists, he expressively works the material, exposing trowel marks and, in some areas, the white plaster.

At first glance, Van Vranken's paintings may be mistaken for three-dimensional reliefs. The abstract planes configured through perspective and value contrasts produce this *trompe l'oeil* effect. The artist also paints and incorporates common signage, like the black running figure stenciled after a sign cautioning motorists of schoolchildren at play. This surreal element is heightened by a sense of order and eternal time, conveyed through the classical, geometric composition.

Combining the techniques and materials of artist and builder, Van Vranken seeks an architecture of balance through which opposing tendencies converge: illusion and abstraction, motion and stasis, order and chaos, traditional and contemporary art. After creating and destroying his foundation of plaster, painting, and finally glazing the work, the artist arrives at this point of synthesis:

> *Inside my paintings, one has the feeling that time has passed, that lives have been lived here, and that things have changed. What remains is something strangely similar to what we see on the abandoned walls of our urban architecture: the tracery of the silent, yearning passage of time. After the storm of time, or in the case of these paintings, the storm of creation has ended, a kind of deep cathartic balance results, contradictions resolve themselves, opposites become one, and this small part of the world is made whole.*

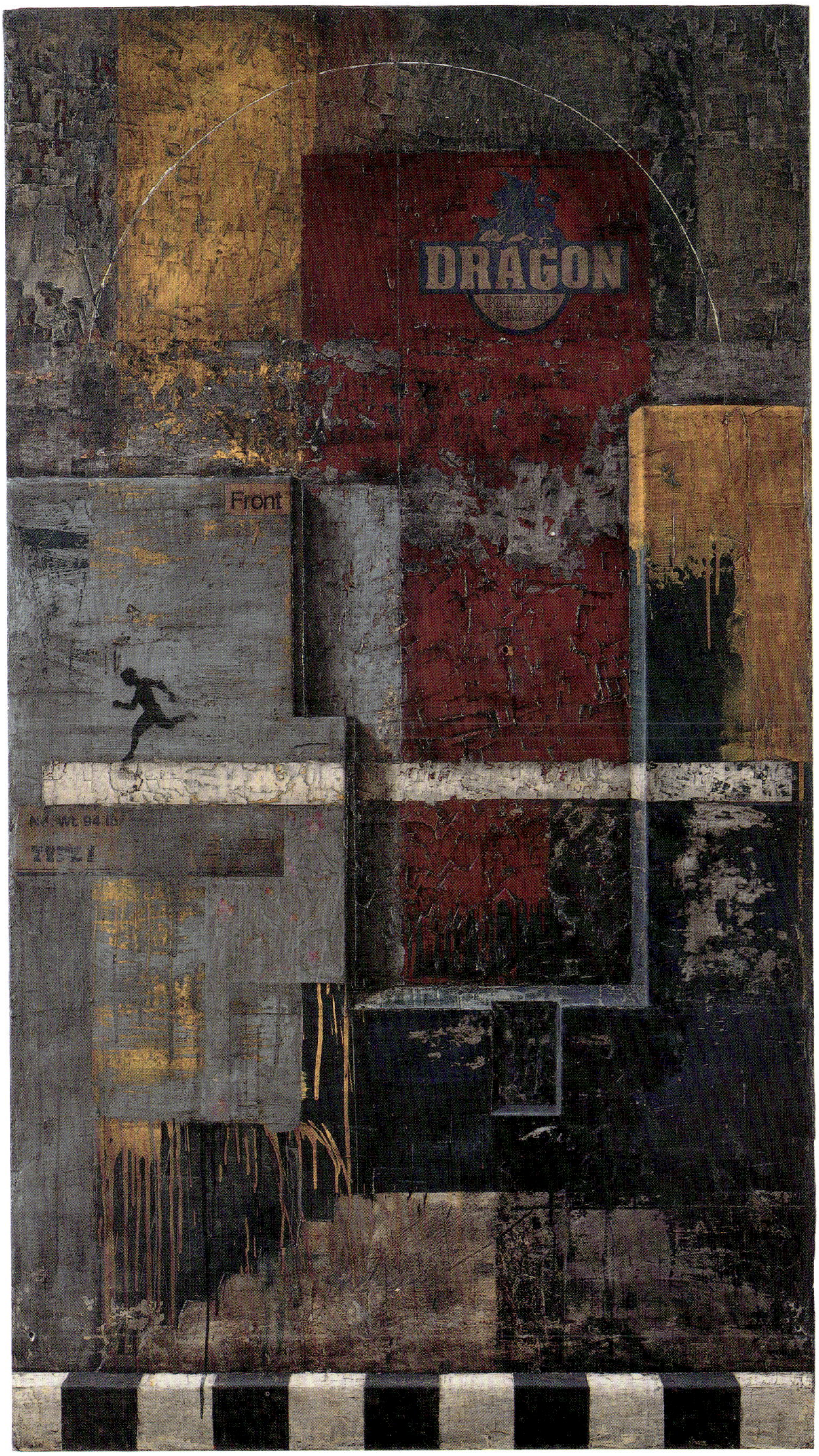

ROBERT VAN VRANKEN
Untitled, 1989
Mixed media on plaster on board
84 x 48 inches

CHRISTOPHER CAIRNS

Christopher Cairns' interest in myth, religion and history led him to reinterpret the traditional female allegorical figure in contemporary terms. For the past five years, the Synagogue figure, a Gothic sculpture from the south portal of Strasbourg Cathedral (1230), has inspired Cairns in several monumental works. This feminine figure's medieval identity as the disenfranchised member of the church is translated into a symbolic reference to victims of the AIDS epidemic in Cairns' own *Synagogue as AIDS Memorial* (1985-90). A still life of hypodermic needles at the sculpture's base refers more graphically to the crisis. In Cairns' monument, ideas of sacrifice and personal heroism span the centuries.

The artist uses plaster to synthesize traditional and innovative views of the female figure. Cairns often models his forms in clay and later casts them in plaster. They are then cut with a knife, sawed, reassembled or temporarily discarded for use in future sculptures. After sections are broken, the artist "mends" them together with a seam of clay, then makes another mold and plaster cast. By this process, arms, legs, and torsos from different figures are often combined. Cairns' process, like Jonathan Silver's, results in a complex assemblage of forms. The effect resembles a Cubist faceting of planes where inside and outside, front and back, can often be viewed simultaneously. Cairns' sculptures also echo the classical tradition; they are deliberately beautiful and seemingly disparate sections always cohere.

Cairns' process is both improvisational and technically precise. Plaster and clay together give the artist the flexibility to resolve a "puzzle" of interlocking convex and concave shards and anatomical parts. Although the artist sometimes casts his work in bronze, he prefers plaster for his figures. Its luminosity makes them seem otherworldly. The cracks that develop during the process are retained, imparting an immediate sense of history. Plaster heightens the vulnerability of Cairns' sculpture and reinforces its dramatic impact.

1942 Born in Wilmington, Delaware

Lives and works in Haverford, Pennsylvania

Professor of Fine Arts, Haverford College

1970 M.F.A. Tulane University, New Orleans, Louisiana

1965 B.A. English Literature, Oberlin College, Ohio

SELECTED AWARDS

1980 Artists Fellowship, Pennsylvania Council on the Arts

1975-1976 Mellon Foundation Grant

1971, 1972 Ford Foundation Grant

SOLO EXHIBITIONS

1988-1989 Morris Arboretum of the University of Pennsylvania, Chestnut Hill, Pennsylvania

1987, 1985 Leslie Cecil Gallery, New York

1985 New York Studio School Gallery, New York

1985 Swarthmore College, Pennsylvania

1984 More Gallery, Philadelphia

1983, 1976 Hamline University, St. Paul, Minnesota

1982, 1979, 1975 Comfort Gallery, Haverford College, Pennsylvania

1977 Washington and Lee University, Lexington, Virginia

1970 Tulane University, New Orleans, Louisiana

SELECTED GROUP EXHIBITIONS

1987 Two Person Exhibition, Haverford College, Haverford, Pennsylvania

1986 "Heads," New York Studio School, New York

1986 "Peter Agostini and His Former Students," Leslie Cecil Gallery, New York

1984 Trenton Museum, New Jersey

1979 Two Person Exhibition with Jonathan Silver, New York Studio School, New York

1978 Weatherspoon Art Gallery, University of North Carolina, Greensboro

1976 Two Person Exhibition with Jonathan Silver, 4x10 Gallery, New York

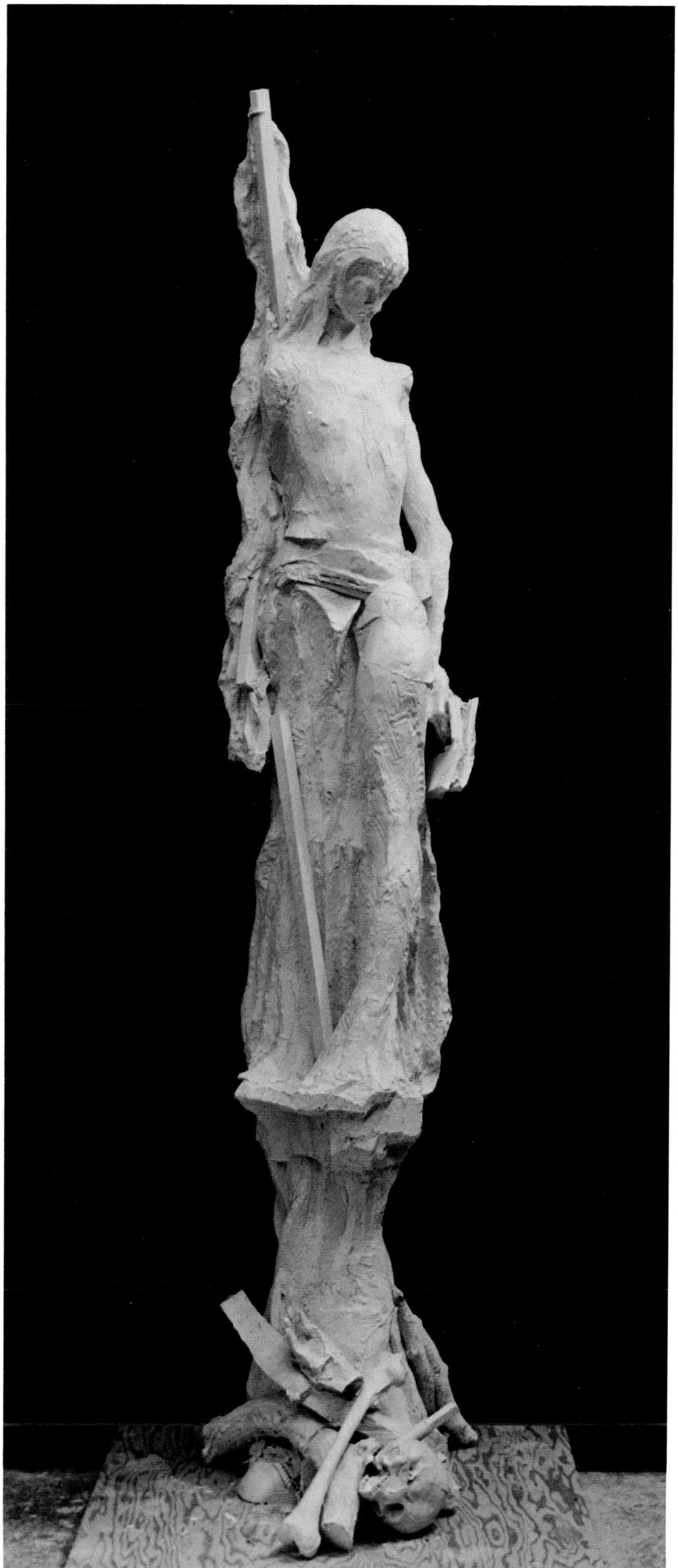

CHRISTOPHER CAIRNS
Synagogue as AIDS Memorial, 1985-90
Plaster
Height: 110 inches

RAY RAPP

1951 Born in Rochester, New York

Lives and works in Brooklyn, New York

1981 M.F.A. San Francisco Art Institute

1977 B.A. Fine Arts, San Francisco State University

1973 B.A. Philosophy, University of California, Berkeley

SELECTED AWARDS

1990 Experimental Television Center Grant, Rotunda Gallery, Brooklyn, New York

1989 Artists Space Artist's Grant, City Gallery, New York

1986 BACA Artist's Grant, BACA Downtown, Brooklyn, New York

1983 Berkshire Museum Exhibition of Painting and Sculpture

SELECTED SOLO EXHIBITIONS

1990 "Enclosure," (installation), Rotunda Gallery, Brooklyn, New York

1989 "Verticality," Bratton Gallery, New York

1986 "Heads I Win, Tails You Lose," (installation), BACA, Brooklyn, New York

1984 "Totem and Tabu," Long Island University, Brooklyn, New York

1980 "Number O," Fresno Community College Gallery, California

SELECTED GROUP EXHIBITIONS

1990 "Image, Object Abstracted," Organization of Independent Artists, New York

1989 "On the Cutting Edge," Fine Arts Museum of Long Island, Hempstead, New York

1988 "Six Views: Installation," City Gallery, New York

1985 "Working in Brooklyn: Sculpture," The Brooklyn Museum, Brooklyn, New York

1984 "Works on Paper by Contemporary Sculptors," The Brooklyn Museum, Brooklyn, New York

1983 "The American Artist as Printmaker," The Brooklyn Museum, Brooklyn, New York

1981 "Selections," Richmond Art Center, Richmond, California

Ray Rapp explores the visual and audio properties of liquid plaster in *Plaster Flow* (1990), a site-specific installation for The Queens Museum. Four video monitors, implanted in free-standing plaster sculptures, offer a continuous, dynamic array of counterpointed sensual patterns. Two screens show plaster pouring from the top of a chute; as it cascades over the edge, different rates of flow translate into rippling rings. Two other monitors feature bubbling, spreading, lava-like plaster as Rapp releases air from balloons and agitates the mixture with a kitchen blender. Throughout the artist's video footage, an element of playfulness underlines the unexpected beauty of his imagery.

In this activated state, accompanied by gurgling and lapping sounds, the plaster evokes a variety of active geological phenomenon. More stable landscape formations — glaciers, mountain escarpments — can also be read into the sculptures that house the video monitors. By contrasting liquid plaster and congealed sculptures with smooth and crystalline edges, Rapp visualizes a process of transformation. In the artist's hands, plaster is a primordial soup, the essential substance from which form emerges.

Although natural forms can be intuited from this installation, the artist also refers to architecture and construction. Rapp developed an interest in plaster three years ago, thinking about walls made of plaster and the idea of enclosure. He conceived of a series of projects that would release the wall from its original function. Based on the principles of concrete building construction, the sculpture is made by pouring the liquid material into a wood frame which is afterwards discarded. When it hardens, the resulting sculptures retain impressions of the wood and the architectural integrity of the form. By contrasting this structural component of plaster with its fluidity, Rapp conceptualizes the material's flexibility and transformational states.

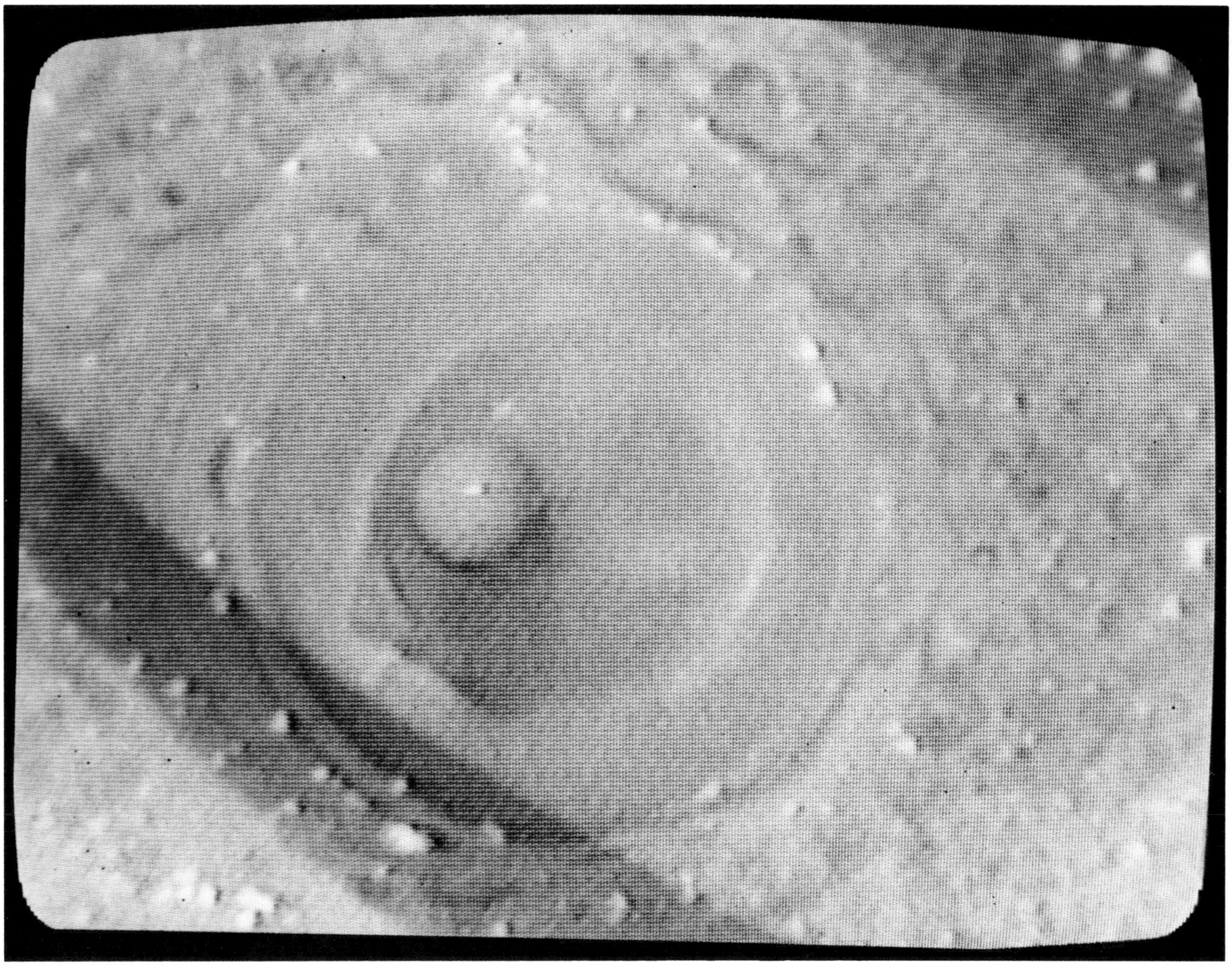

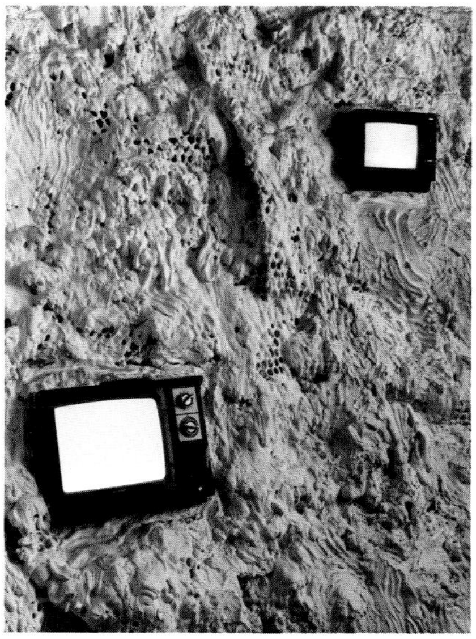

RAY RAPP
Video image from *Plaster Flow*, 1989-90
Site-specific installation
for the Lobby Gallery
Plaster, joint compound, plywood, wire,
four video monitors
Three sculptures: 96 x 120 x 96 inches
96 x 6 x 37 inches, 96 x 38 x 8 inches

Detail from *Enclosure*, 1990
Rotunda Gallery, Brooklyn

ELFI SCHUSELKA

1940 Born in Vienna, Austria

Lives and works in New York City

1959-1961 Academy of Applied Fine Arts, Vienna

1958-1960 Art History & Theatre, University of Vienna

1958-1959 Photography, Graphic & Experimental Institute, Vienna

SELECTED AWARDS

1988 Artist's Space Grant

1986 New York Foundation for the Arts (Painting)

SELECTED SOLO EXHIBITIONS

1988 New York University Broadway Windows, New York

1988 Neue Gallery, Vienna, Austria

1987, 1985, 1983, 1980 Condeso/Lawler Gallery, New York

1986 American Center, Belgrade, Yugoslavia (travelling exhibition)

1985 Joan Hodgell Gallery, Sarasota, Florida

SELECTED GROUP EXHIBITIONS

1989 "International Biennale," Varna, Bulgaria

1989 "International Biennial," Bharat Bhavan, India

1988-1989 Bokushin Gallery, Tokyo, Japan

1987-1988 "Big Drawings," Univer-sity of Colorado at Colorado Springs; University of Wyoming Art Museum at Laramie

1986 "Monumental Drawing," The Brooklyn Museum, New York

1985 "Fête de la Gravure," Musée d'Art Moderne, Liege, Belgium

1983 Fundacion Joan Miro, Barcelona, Spain

1982 Edith C. Blum Art Institute, Bard College, Annandale-on-Hudson, New York

1984, 1982, 1980 Kyoto Municipal Museum of Art, Kyoto, Japan

1980 "Between Object and Illusion," Rutgers University, New Jersey

1978 "Contemporary Reflections," Aldrich Museum of Contemporary Art, Connecticut

1974 University of Dallas, Texas

Elfi Schuselka's artistic process is her vehicle for the cathartic release of life's pains and frustrations. In the spirit of the Abstract Expressionist artists, she defines her existence through the very act of making art; however, Schuselka expands the parameters of "action painting" by demolishing her own work. Paradoxically, in the aftermath of violence, a shimmering, delicate veil of plaster in relief is revealed. As in the eye of a hurricane, a strange, meditative calm prevails.

Elfi Schuselka began working with plaster in the 1960s. By 1977, she had discovered a unique process to create relief panels. After weaving a jute net, the artist coats the fibers with plaster. She paints one side of the surface and then proceeds to smash it with a hammer. The act of shattering the plaster results in an intricate web of shards as well as a mound of "fallout" on the floor which she originally exhibited. In these works, subtle passages of pastel color are randomly interspersed with raw plaster fragments of the unpainted side reversed from the force of hammering. Schuselka creates an all-over rhythm of solids and voids as the surviving chunks of plaster intervene between vacant spaces. Cracked plaster edges and frayed jute emphasize the emotional and physical states of fragility and ruin.

The plaster wall is Schuselka's point of departure. Living on the Lower East Side among abused buildings and interiors, the artist finds their ambiguity, impermanence and adaptability compelling. She captures this state of flux by using plaster and tempting chance. Although Schuselka accepts random variables and accident in both art and life, she nevertheless imposes order: the jute net, based on the minimalist grid structures her reliefs. In these works, the physical and emotional states of chaos and balance, destruction and peace, are resolved.

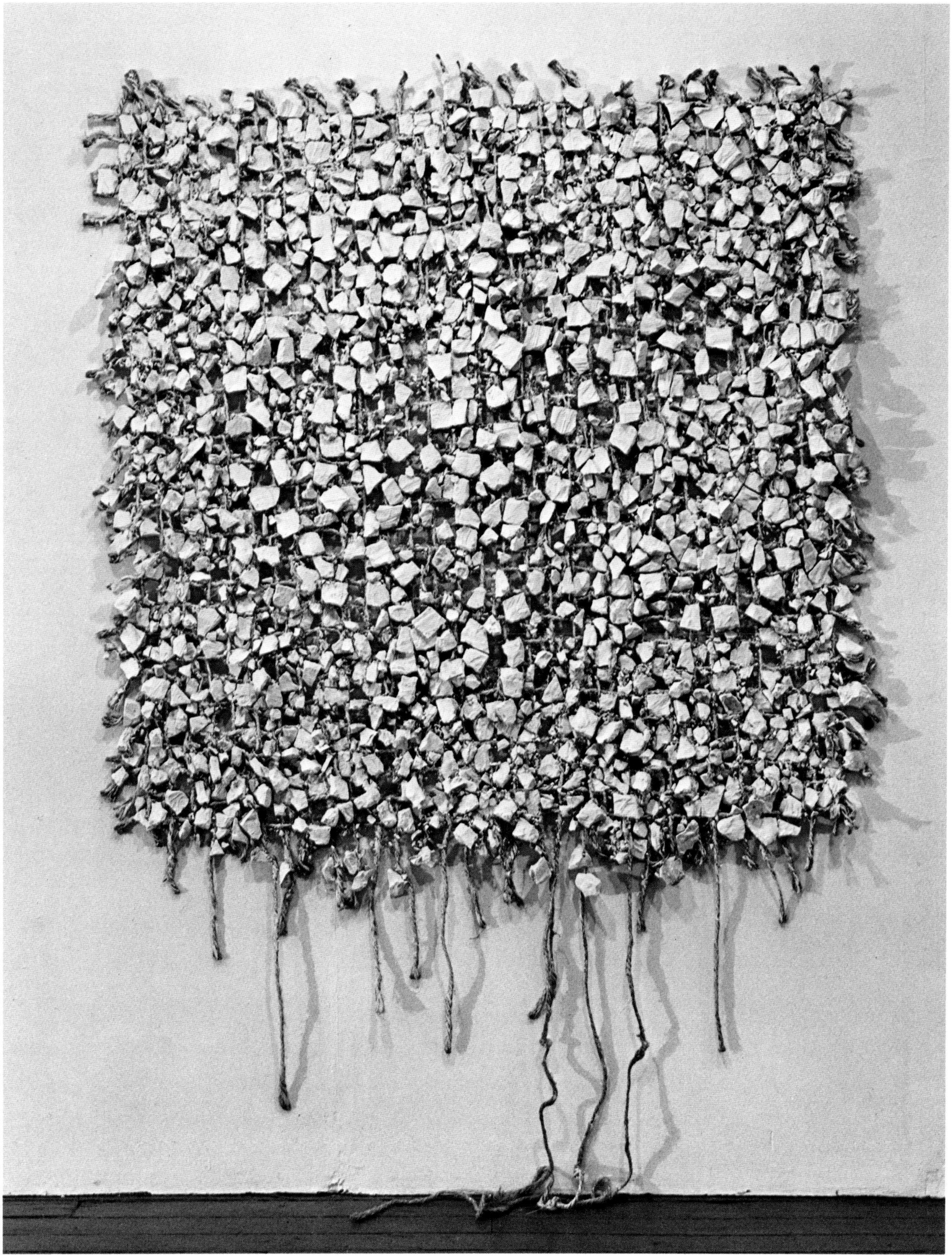

ELFI SCHUSELKA
Wall piece #25, 1978
Plaster, acrylic, jute, twine, monofilament
87 x 64 inches

KATIE SEIDEN

1946 Born in New York City

Lives and works in Sea Cliff, New York

SELECTED AWARDS

1988 Artists Grant, Artists Space, New York

1978 Award of Excellence, Heckscher Museum, Huntington, New York

SELECTED SOLO EXHIBITIONS

1988 Washington County Museum of Fine Arts, Hagerstown, Maryland

1988, 1987 Sensory Evolution Gallery, New York

SELECTED GROUP EXHIBITIONS

1989-90 "Lines of Vision: Drawings by Contemporary Women," Islip Art Museum, East Islip, New York. Travelling to Fine Arts Museum of the South, Mobile, Alabama; Chattanooga Regional History Museum, Chattanooga, Tennessee; Longview Museum of Art, Longview, Texas

1989 "One Hundred Women's Drawings," Hillwood Art Gallery, Brookville, New York and Blum Helman Warehouse, New York

1989 University of Hawaii Art Gallery, Honolulu, Hawaii

1988 San Francisco International Airport, San Francisco, California

1986 "Letters," The Clocktower, New York

1986 "Art of The Northeast USA," curated by Linda Shearer, Silvermine Guild Galleries, New Canaan, Connecticut

1984, 1980, 1979 Del Mar Gallery, Corpus Christi, Texas

1982 "The Wild Show," curated by Faith Ringgold, Project Studios I, Long Island City, New York

1981 "Kindred Spirits: Works by Joseph Cornell and others," Heckscher Museum, Huntington, New York

Using classical myths as her point of departure, Katie Seiden creates anomalous sculptures by encasing mundane objects, like shoes and boots, in plaster. Seiden's attitude toward the object updates the tradition of early twentieth-century Dada and Surrealist artists who infused mystery and ambiguity into ordinary things by placing them in unexpected contexts.

Seiden's personal transformation of materials parallels her fascination with the myth of the wood nymph, Daphne, who is metamorphosed into a tree to escape Apollo's pursuit. In *Apollo and Daphne V* (1988), the final sculpture in a five-year series, Apollo seems to squeeze the life out of Daphne. Reduced to a columnar trunk, she is constricted by a twining skein of plaster embedded with spiked-heeled shoes. Symbols of feminine seduction as well as weapons of defense, the spiked heels heighten the message of sexual violation. Seiden's objects subtly retain their identity as they coalesce into the fabric of her work. They also lend color to the plaster where they are exposed.

This sculpture is very baroque in effect — a spiraling composition with deep, dark shadows collecting in holes. Through billowing forms of plaster, Seiden distills a transitory moment of emergence and disappearance. Compared to Bernini's classic Baroque sculpture of *Apollo and Daphne* (1622-25) where the encounter between male and female mythological beings is interpreted so delicately and harmoniously, Seiden's work is clearly confrontational. Both sculptors express the energy of movement inherent in the subject. While Bernini illusionistically carves his marble sculpture into representational figures, Seiden translates the myth abstractly and in pointedly feminist terms.

Seiden has been working with plaster for ten years. Starting as a painter, she turned to plaster sculpture because, apart from studio casting, there was no tradition behind it. The technical challenges, the material's rough texture, and the potential for dramatic shadows and contrasts attracted the artist. The fact that it is an elusive material, hardens quickly and yellows over time (peculiar to hydrostone alone) reinforces the ambiguity of permanence and change at the core of Seiden's sculptures.

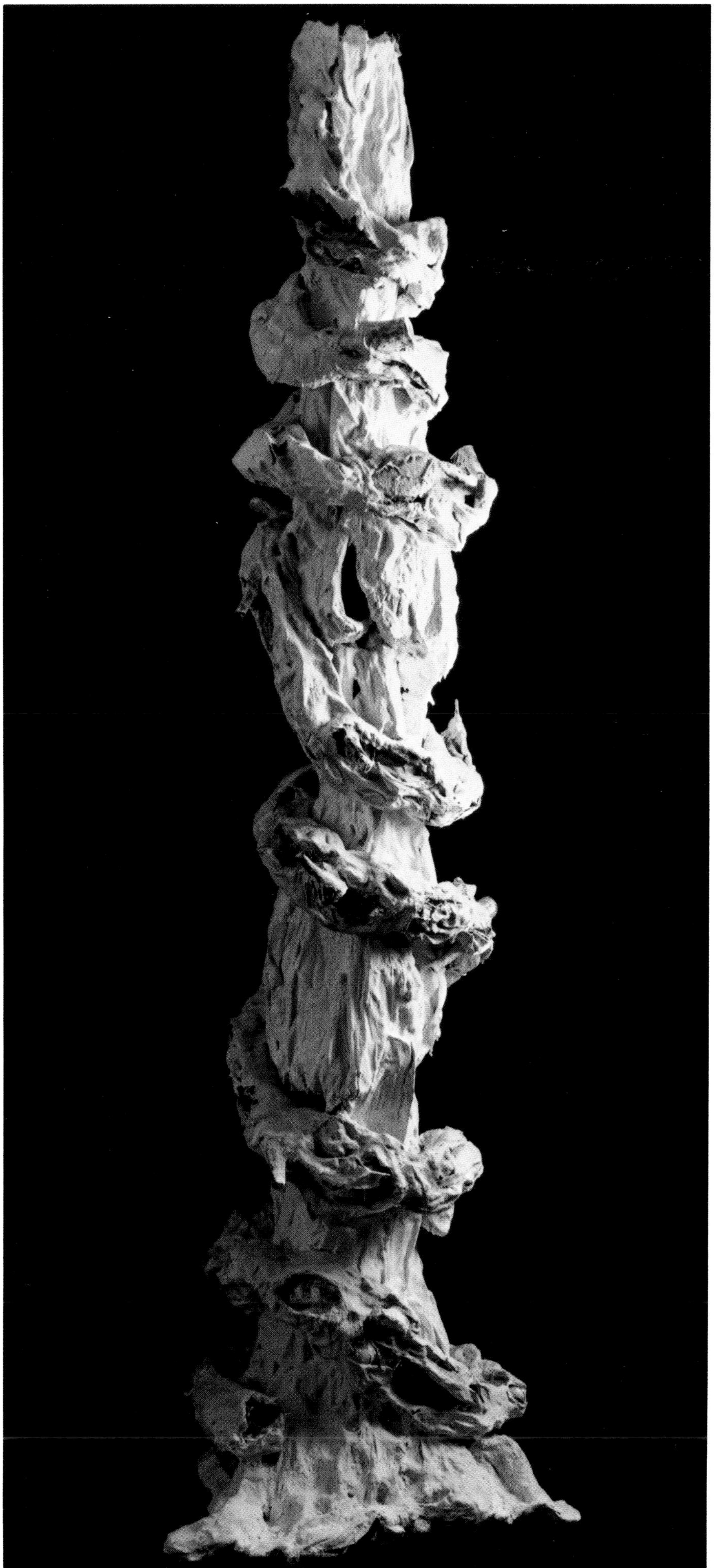

KATE SEIDEN
Daphne and Apollo V, 1988
Hydrostone and encased objects
87 x 24 x 26 inches

CHECKLIST

* Reproduced in the catalogue

PETER AGOSTINI

1. *Figure Turning into a Tree*, 1956
Hydrocal and burlap
42 x 15 x 4 inches
Courtesy of the Artist, New York

2. *Summer Breezes*, 1962*
Hydrocal and burlap
3 sculptures: 36 x 30 x 18 inches
36 x 20 x 20 inches
36 x 28 x 20 inches
Courtesy of the Artist, New York
Photograph: Don Cook

3. *The Wave*, 1963*
Hydrocal, chicken wire and cheesecloth
62 x 44 x 22 inches
Courtesy of the Artist, New York
Photograph: Sarah Wells

4. *Ghost Horse*, 1986
Hydrocal and burlap
47 x 40 x 18 1/2 inches
Courtesy of the Artist, New York

CHRISTOPHER CAIRNS

5. *Synagogue as AIDS Memorial*, 1985-90*
Plaster
Height: 110 inches
Courtesy of the Artist, Haverford, Pennsylvania

6. *Black Madonna*, 1984
Plaster
Height: 94 inches
Courtesy of the Artist, Haverford, Pennsylvania

CLAUDE FIXLER

7. *Putting Ourselves to Bed*, 1989-90*
Site-specific installation for the West Gallery
Hydrocal, plaster gauze, wood
(Drawing: ink wash and graphite on paper
22 x 15 inches)
Photograph: Lynn Whitney

ANITA CURTIS GLESTA

8. *Seated Figure*, 1987*
Plaster and mixed media
70 x 56 inches
Courtesy of the Artist, Brooklyn, New York
Photograph: Sarah Wells

9. *Untitled*, 1989
Plaster and mixed media
44 x 22 inches
Courtesy of the Artist, Brooklyn, New York

BARRY GOLDBERG

10. *Heartwood*, 1987
Hydrocal, steel and mixed media
109 x 13 1/2 x 13 1/4 inches
Courtesy of the Artist, New York

11. *Drifting Buoy*, 1989*
Hydrocal, steel and mixed media
100 x 36 x 12 inches
Courtesy of Cavaliero Fine Arts, New York
Photograph: Larry Wheelock

GRACE KNOWLTON

12. *Sphere*, 1989
Design Cast and mixed media
Diameter: 66 inches
Courtesy of Bill Bace Gallery, New York

13. *Shard*, 1988
Painted Design Cast
78 x 90 x 1 inches
Courtesy of Bill Bace Gallery, New York

14. *Triptych*, 1989*
Design Cast, clay, mud, iron pigment
Each panel: 96 x 24 x 1/2 inches
Courtesy of Bill Bace Gallery, New York
Photograph: Gordon Rapp

SUSAN LAUFER

15. *Deep Blue*, 1987-1990*
Acrylic, polymer, and plaster on panels
108 x 134 inches
Courtesy of Germans Van Eck Gallery, New York
Photograph: Erik Landsberg

16. *Beacon*, 1988-89
Acrylic, polymer, and plaster on panels
99 x 84 inches
Courtesy of Germans Van Eck Gallery, New York

MANUEL NERI

17. *Untitled Male Head*, 1959
Plaster, enamel, graphite
9 x 6 x 4 inches
Courtesy of the Artist and
Charles Cowles Gallery, New York

18. *Shrouded Figure*, 1961*
Plaster, pigment, aluminum, burlap
28 x 32 x 15 inches
Collection of Paule Anglim, San Francisco

19. *Seated Girl*, 1964*
Plaster, enamel, wood
44 x 29 1/2 x 27 1/2 inches
Courtesy of Mr. & Mrs. Robert Hudson,
Cotati, California
Photograph: M. Lee Fatherree

20. *Catun Series I*, 1986-88
Plaster, dry pigment, string, steel and wood
69 1/2 x 22 x 13 1/2 inches
Courtesy of Charles Cowles Gallery, New York